# Fred's Home Companion

# Advanced Algebra

*Stan Schmidt*

*Fred's Home Companion*
*Advanced Algebra*

Stanley F. Schmidt, Ph.D.

Polka Dot Publishing

ISBN-13: 978-0-9709995-7-3
ISBN-10: 0-9709995-7-7
Library of Congress Catalog Number: 2005905757
Printed and bound in the United States of America


Polka Dot Publishing        P. O. Box 8458        Reno NV 89507-8458

*For full information regarding price and availability of this book, e-mail Polka Dot Publishing at:* lifeoffred@yahoo.com

PolkaDotPublishing.com


Illustrated by the author with additional clip art furnished under license from Nova Development Corporation, which holds the copyright to that art.

*for Goodness' sake*

or as J.S. Bach—who was
never noted for his plain
English—often expressed it:

*Ad Majorem Dei Gloriam*
(to the greater glory of God)

If you happen to spot an error that the author, the publisher, and the printer missed, please let us know with an e-mail to: lifeoffred@yahoo.com

SPECIAL OFFER

As a reward, we'll e-mail back to you a list of all the corrections that readers have reported.

# What is Fred's Home Companion?

It is lots of things. Since *Life of Fred: Advanced Algebra* was first published, there have been requests from home schoolers, teachers, and adults who are learning advanced algebra. This book is a response to those needs.

Need #1: I'm a home schooler and I would like my algebra chopped up into daily bite-sized pieces.

Done! *Fred's Home Companion Advanced Algebra* offers you 101 daily readings. In the space of one summer, for example, you can finish all of advanced algebra and still have plenty of time to do other things.

Need #2: I'm a classroom teacher and I want someone (like you!) to write out all my lesson plans and lecture notes.

Done! Here are 101 lesson plans. Each one tells you what part of the book you'll be covering. Lots of the lessons include problems which are not in *LOF:AA*. These can serve as your lecture notes. Or as material for pop quizzes or for tests. The answers are also supplied to these problems.

Need #3: I'm an adult working my way through your LOF:AA book and I'd like the answer key for the end-of-the-chapter problems. In the book you give the answers to half the problems.

Done! Here is the answer key.

Need #4: I'm in need of a lot of practice in algebra. Although you have a lot of problems for me to work on in LOF:AA, I want a bunch more. I really want to pound it into my head.

Done! In this book we supply a ton of additional advanced algebra problems. Finish all of these problems in addition to the ones in *LOF:AA,* and you should be able to join Fred as a professor of mathematics at KITTENS University.

# A Note to Students

When you turn to Lesson One in this book, you will find that it asks you to read three pages in *Life of Fred: Advanced Algebra*. Reading a little bit about Fred and his adventures is always a fun way to begin a day. In the first lesson you'll be reading about Fred heading home on the bus. He's wearing his hospital shirt with little blue and green frogs on it that he received in *Life of Fred: Beginning Algebra*.

After you have read those three pages, turn back to this book and answer the questions of Lesson One. All the answers are given on the next page, so you'll know you are on the right track.

That's it.

Ahead of you are
   *Life of Fred: Geometry*
   *Life of Fred: Trig*
   *Life of Fred: Calculus.*

*Life of Fred: Geometry* is really a book about reasoning. The circles and triangles and squares and lines and points and rectangles and planes and parallelograms and polygons and angles just give us something to reason about. You will learn about what makes a valid argument. This is the only mathematics course from kindergarten through the second year of college that concentrates on what it means to think logically. You encounter lots of *If-Then* statements.

Suppose your mother, speaking hyperbolically, tells you, "If you do *that*, I'm going to kill you." And suppose you do *that*. You might not be ready for geometry. Or, on the other hand, geometry may be exactly what you need.

*Life of Fred: Trig* deals mostly with triangles. You might be given

this right triangle 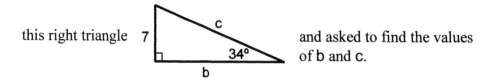 and asked to find the values of b and c.

Or in trig you might be
given this triangle
and asked to find c.

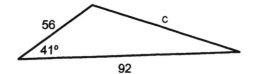

Calculus is the gateway to many different majors in college. All the sciences need it. Even business majors often need it. You may not need calculus if all you are going to take is It Lit.*

Calculus adds one new concept—that of the idea of limit. The definition of limit is the wellspring from which all of the three or four semesters of calculus flow.

Here are three of the many uses of the idea of limit:

Suppose we have the curve $y = -x^2 + 8x - 12$.

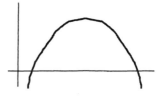

Use #1: We can find the maximum point on the curve.

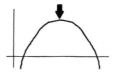

Use #2: We can find the area under the curve which is above the x-axis.

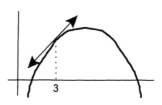

Use #3: We can find the slope of the line that is tangent to the curve at $x = 3$.

———————————

* "It Lit" is college-talk for Italian literature courses.

Now to answer some of the common questions that students have . . .

## CAN I USE MY CALCULATOR IN ALGEBRA?

Yes. It is the addition and multiplication tables that you need to know by heart. Once you have them down cold, and you know that the area of a triangle is one-half times base times height, there is little else that you should have to sit down and memorize.

When I taught arithmetic, the tests I gave were always taken without the use of a calculator, but when I taught algebra/geometry/ trigonometry/calculus/math for business majors/statistics, the tests were always open-book, open-notes, and use-a-calculator-if-you-want-to.

There are a lot of times in life in which you may need to know your addition and multiplication facts and won't have access to a calculator, but when you are doing algebra or calculus problems you will almost always have a calculator and reference books handy.

## WHAT KIND OF CALCULATOR WOULD BE GOOD?

In beginning algebra all you really needed was the basic calculator that has these five keys: $+, -, \times, \div, \sqrt{\phantom{x}}$ .

Now in advanced algebra it is time to buy a "scientific calculator." It will have sin, cos, tan, !, log, and ln keys. The most fun key is the "!" key. If you press 8 and then hit the ! key, it will tell you what $8 \times 7 \times 6 \times 5 \times 4 \times 3 \times 2 \times 1$ is equal to.

Recently, I saw a scientific calculator on sale for less than $8. This will be the last calculator you will need to learn all the stuff through calculus.[*]

---

[*] Some schools require their calculus students to buy a fancy graphing calculator which costs between $80 and $100. I don't own one and I've never needed one. I spent the money I saved on pizza.

# A Note to Parents Who Are Homeschooling Their Kids

Fred's Home Companion will put your children on "automatic pilot." Each day they do one (or more) lessons. The reading in *Life of Fred: Advanced Algebra* is fun. And because it is fun, they will learn mathematics much more easily. You can sit back and watch them learn.

Six-year-old Fred first encounters the need for mathematics in his everyday life, and then we do the math. This is true for all the *Life of Fred* books. The math is *relevant*.

In the traditional school settings, all the subjects are packaged into air-tight compartments. The students are like little cars that scurry around during the day. First, they might park themselves in the English "filling station" and get a gallon of English poured into them. At the history filling station they would get history.

For years, educators have bemoaned this compartmentalization. "We are teaching children, not subjects," is a favorite expression of theirs. And yet, things don't seem to change much. If it is nine o'clock and the students are sitting in a French class, they never hear about biology or music. This is not a natural way to learn. Do French people only discuss how irregular their verbs are?

In *Life of Fred: Advanced Algebra* we certainly teach algebra—more than is taught in most schoolrooms—but the whole world is ours to explore and learn about.

In chapter four, when the four-year-old girls eat some doughnut dough, we discuss the **biology**, **chemistry**, and **physics** (p. 117). When they float up in the sky, they seem to take the shape of the constellation Cassiopeia. We do a little **astronomy** and mention how many official constellations there are (p. 124). When the five little girls land in a tree (p. 127), it is time to quote a little **poetry**: ". . . A tree that looks at God all day/And lifts her leafy arms to pray. . . ." which was written by Joyce Kilmer who "died in action on July 30, 1918, near the end of WWI." In the space of two sentences we have mentioned **prayer** in a positive context, **patriotism**, and that World War I ended in 1918 (**history**).

In those same eleven pages (pp. 117–127) we have also done a lot of algebra:

✼ What a line with slope zero looks like,

✼ What a line with a negative slope looks like,

✼ A whole-page argument as to why we use subscripts,

✼ The Greek letters sigma and pi,

✼ The derivation of the formula for the slope of line passing through the points $(x_1, y_1)$ and $(x_2, y_2)$,

✼ The derivation of the distance formula between the points $(x_1, y_1)$ and $(x_2, y_2)$,

✼ The slope-intercept and the double-intercept forms of the line,

✼ The point-slope form of the line,

✼ The concepts of independent and dependent variables, and

✼ The derivation of the two-point form of the line.

A lot happens in *Life of Fred: Advanced Algebra*—a tale of Fred's two-day bus trip from Texas to Kansas.

# Contents

# *Lesson One*

## Ratios, Median Averages, Proportions

*Life of Fred:*
*Advanced Algebra*
pp. 16–18

1. Which is larger: 6:5 or 9:8?

2. In some of the old math books they used to write a proportion as 2:3::6:9. What would the double colon in the middle represent?

3. When Fred first counted the ratio of passing telephone poles to his heartbeats, he found it was 5:3. Suppose the driver of the bus increased his speed. What might the new ratio look like?

4. As Fred was counting the ratio of passing telephone poles to his heartbeats, suppose (Heaven forbid!) his heart stopped beating.

    ✓ The bus driver wouldn't like this because he would have to stop the bus and do some heart surgery or something.

    ✓ The readers of the advanced algebra book wouldn't like it because the book would end too soon.

    ✓ Mathematicians wouldn't like it because the resulting ratio is 5:0. Why would they object?

5. Solve $\dfrac{x+3}{x+13} = \dfrac{3}{5}$

6. The bus driver is 25 years old. The bus is 35 years old. How long will it be before the driver is 75% of the age of the bus?

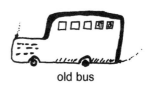

old bus

7. What is the median average of:

    5, 8, 9, 9, 10, 14, 18, 19, 19?

1. 6:5 means $6 \div 5$ which is 1.2.
    9:8 means $9 \div 8$ which is 1.125.      6:5 is larger.
2. A proportion is the equality of two ratios.
    The expression 2:3::6:9 would translate into 2:3 = 6:9  or  $\dfrac{2}{3} = \dfrac{6}{9}$

3. Instead of 5:3 it might be 6:3 or 7:3.  Any answer you gave which was in the form x:3 where x > 5 would have been fine.

4. A ratio of 5:0 means $\dfrac{5}{0}$ which is division by zero.  Mathematicians don't especially like that.  It is similar to going up to someone and saying, "The snamplefork is overzipped."  Division by zero *doesn't have any meaning*.  When you divide 2 into 6 you get an answer of 3.

$$2\overline{)6}^{\,3}$$

You check your answer by multiplying 2 by 3 and hoping to get 6.

If you try to divide by zero,   $0\overline{)6}^{\,?}$   what could the answer be?  What number could you replace the question mark with so that the answer would check?  Suppose the answer were 97426398799426.

Suppose      $0\overline{)6}^{\,97426398799426}$

This answer wouldn't check since $0 \times 97426398799426 \neq 6$.

5.          $\dfrac{x+3}{x+13} = \dfrac{3}{5}$

$\dfrac{(x+3)5(x+13)}{x+13} = \dfrac{3 \cdot 5(x+13)}{5}$      Multiplying both sides by **5(x + 13)**

$(x+3)5 = 3(x+13)$

$x = 12$

6. Let x = the years until the bus driver is 75% of the age of the bus.
    In x years, the bus driver will be 25 + x years old.
    In x years, the bus will be 35 + x years old.
    $75\% = \dfrac{3}{4}$

$\dfrac{25+x}{35+x} = \dfrac{3}{4}$

$x = 5$ years

7. The median average of 5, 8, 9, 9, 10, 14, 18, 19, 19 is the number in the middle when they are all arranged in order of size.  In this case it is 10.

$\mathcal{L}esson\ \mathcal{T}wo$

## Solving Proportions by Cross-Multiplying

1. Solve by cross-multiplying:

$$\frac{x-4}{x+2} = \frac{1}{3}$$

*Life of Fred:*
*Advanced Algebra*
pp. 19–21

2. Cheryl Mittens is twice as old as the watch she wears. Six years ago (when Fred was born) she was three times older than her watch. How old is her watch now?

3. Solve $\dfrac{x-4}{2} = \dfrac{5}{x+5}$   (Use factoring to solve the resulting quadratic equation.)

4. Solve $\dfrac{x}{x^2-10} = \dfrac{5}{3}$   (Use the quadratic formula.)

Cheryl's watch that she got when she was in middle school

5. Cross-multiplying works fine with proportions.
$\dfrac{5}{\pi} = \dfrac{x}{11}$   turns into  $55 = \pi x$.

But when it's not a proportion, we use other approaches.

What would be the *next step* in solving   $\dfrac{5}{\pi} = \dfrac{x}{11} + \dfrac{x-3}{x}$

1.
$$\frac{x-4}{x+2} = \frac{1}{3}$$

$$3(x-4) = x+2$$
$$3x-12 = x+2$$
$$2x = 14$$
$$x = 7$$

2.  Let x = the current age of her watch.
    Then 2x = Cheryl's current age.
    Six years ago her watch was x – 6 years old.
    Six years ago she was 2x – 6 years old.

$$\frac{x-6}{2x-6} = \frac{1}{3}$$

$$x = 12 \qquad \text{Her watch is 12 years old.}$$

3.
$$\frac{x-4}{2} = \frac{5}{x+5}$$
$$x^2 + x - 20 = 10$$
$$x^2 + x - 30 = 0$$
$$(x+6)(x-5) = 0$$
$$x + 6 = 0 \ \text{OR} \ x - 5 = 0$$
$$x = -6, 5$$

4.
$$\frac{x}{x^2 - 10} = \frac{5}{3}$$
$$5x^2 - 3x - 50 = 0$$
$$x = \frac{3 \pm \sqrt{9 + 1000}}{10}$$

5. Actually, there are two possible next steps for $\dfrac{5}{\pi} = \dfrac{x}{11} + \dfrac{x-3}{x}$

One way is to add the two fractions on the right and then you would have a proportion in which you could cross-multiply.

A second way would be to solve this as a fractional equation. You would multiply each term by an expression that all the denominators would divide evenly into:   $\dfrac{5 \cdot 11\pi x}{\pi} = \dfrac{x \cdot 11\pi x}{11} + \dfrac{(x-3)11\pi x}{x}$

## *Lesson Three*

### Constants of Proportionality

---

Do the *Your Turn to Play* on pp. 21–23.

> *Life of Fred:*
> *Advanced Algebra*
> pp. 21–23

---

## *Lesson Four*

### Inverse Variation

---

Do the *Your Turn to Play* on p. 26.
Do the *Your Turn to Play* on p. 28–29.

> *Life of Fred:*
> *Advanced Algebra*
> pp. 24–28

---

## *Lesson Five*

### End of the Chapter—Review & Testing
### Part One

---

> *Life of Fred:*
> *Advanced Algebra*
> Read pp. 29–33
> to see how variation might
> be used in biology
> and then
> tackle the first two
> Cities starting
> on p. 34

Do all the problems in the first two cities.

 Eldred

Panora

*Lesson Six*

End of the Chapter—Review & Testing<br>Part Two

---

Do all the problems in
the second pair of cities.

 Talco

Odd answers are in
the text, and even
answers are
given here.  Wagon Mound

*Life of Fred:*
*Advanced Algebra*

The Cities starting
on p. 35

---

### ✦ EVEN ✦ ✦ ANSWERS ✦

Talco

2. k = 4000       4. There's no difference.       6. 343 lbs.

Wagon Mound

2. 19 years       4. No. Hooke's law, where the amount of stretch is
proportional to the amount of weight on the spring, only works within the
spring's normal elastic limits. You can break the spring or stretch it out of
shape by putting too much weight on it. (This problem is very similar to
problem 3 of Panora.)

6. The amount of air you expend varies directly as the number of seconds
you hold the note.     A = ks

---

*Lesson Seven*

End of the Chapter—Review & Testing<br>Part Three

---

Do all the problems in
the third pair of cities.

*Life of Fred:*
*Advanced Algebra*

The Cities starting
on p. 36

(Answers on next page.)     Yellville

Ida

Yellville

1. $f = k\sqrt{t}$     2. $k = 130$     3. proportion     4. $x = 5$

5. 1875 ounces which is 117 pounds (rounding to the nearest pound)

6. The total number of units you earn varies directly as the number of semesters you complete.    $E = ks$

Ida

1. $d = kt^2$     2. 400 feet     3. 110 years old     4. $x = 1/3$

5. The natural numbers $\{1, 2, 3, 4, \ldots\}$ wouldn't be a good set since it doesn't contain zero which is a very common number of jelly beans found in people's pockets.    6. $k = 2.4$     7. $R = 96$     8. 216 lbs.

# *Lesson Eight*

$$x^a x^b = x^{a+b}, \ x^a/x^b = x^{a-b}, \ (x^a)^b = x^{ab}, \ x^{-a} = 1/x^a, \ x^0 = 1, \ (xy)^a = x^a y^a, \ x^{\frac{1}{2}} = \sqrt{x}$$

Do all the problems on these two pages.

Odd answers are in the text, and even answers are given here. ⬛➤

```
┌─────────────────────┐
│   Life of Fred:     │
│  Advanced Algebra   │
│     pp. 39–40       │
└─────────────────────┘
```

2. $d^{-37}$   4. $x$   6. $y^6$   8. $g^{-7}$   10. $a^{1.6}$   12. $x^{440}$   14. $z^{55}$   16. $r^{m-7}$
18. $m^{n+4}$   20. $1/27$   22. $1$    24. $w$   26. undefined    28. $1 \ (b \neq 0)$
30. $1/c^5$   32. $w^4 z^{28}$   34. $w^{49}/r^7$   36. $1$    38. $2$    40. $1/3$    42. $1$

# *Lesson Nine*

$$\sqrt{x}\sqrt{y} = \sqrt{xy}, \text{ Rationalizing the Denominator}$$

Do all the problems on these two pages.

Odd answers are in the text, and even answers are given on the next page.

```
┌─────────────────────┐
│   Life of Fred:     │
│  Advanced Algebra   │
│     pp. 41–42       │
└─────────────────────┘
```

44. 10  46. $2\sqrt{2}$  48. $a^{10}c^4\sqrt{b}$  50. $\sqrt[4]{b}$  52. $2\sqrt{3}$  54. $6\sqrt{y}\ /\ y$
56. $2\sqrt{c}\ /\ c$  58. $(30 - 6\sqrt{w}\ )/(25 - w)$  60. $(88 + 8\sqrt{cd}\ )/(121 - cd)$
62. $(m\sqrt{m} - m\sqrt{n}\ - n\sqrt{m}\ + n\sqrt{n}\ )/(m - n)$

---

# *Lesson Ten*

## Solving Radical Equations, Extraneous Roots

Do the *Your Turn to Play*.

> *Life of Fred:*
> *Advanced Algebra*
> pp. 43–47

---

# *Lesson Eleven*

## Surface Area of a Cone

Do the *Your Turn to Play*.

> *Life of Fred:*
> *Advanced Algebra*
> pp. 48–51

---

# *Lesson Twelve*

## The Complete History of Mathematics (almost)

This twelve-page novel is dedicated to all history majors. They will recognize the two major themes which flow through every well-written chronicle of mankind. First, our knowledge has increased over the years. Second, we haven't changed a bit.

> *Life of Fred:*
> *Advanced Algebra*
> pp. 52–63

# *Lesson Thirteen*

## The Mass of Moving Objects or A Ten-Ton Nickel

Do the *Your Turn to Play.*

*Life of Fred:*
*Advanced Algebra*
pp. 63–66

# *Lesson Fourteen*

## End of the Chapter—Review & Testing
## Part One

Do all the problems in the first two cities.

*Life of Fred:*
*Advanced Algebra*

The Cities starting
on p. 67

 Elgin

 Raleigh

# *Lesson Fifteen*

## End of the Chapter—Review & Testing
## Part Two

Do all the problems in
the second pair of cities.

*Life of Fred:*
*Advanced Algebra*

The Cities starting
on p. 67

Odd answers are in
the text, and even
answers are
given here.

 Valders

 Yellow Pine

## ✸EVEN✸ ✸ANSWERS✸

Valders

2. no solution    4. $\sqrt[20]{abcd}$    or $(abcd)^{1/20}$

6. $x = \pm\sqrt{10}$    8. i

Yellow Pine

2. $2w^{50}\sqrt{3w}$     4. $x = -3$

6. This was a proof that $\sqrt{3}$ is not a rational number.

8. $7 - i$

---

## *Lesson Sixteen*

### End of the Chapter—Review & Testing
### Part Three

Do all the problems in
the third pair of cities.

*Life of Fred:*
*Advanced Algebra*

The Cities starting
on p. 69

 Dagmar

Imperial

---

## ✸A✸N✸S✸W✸E✸R✸S✸

**Dagmar**

1. $w = 3$     2. $12xy^5\sqrt{y}$

3. $4y^{10}/x^6$     4. $(4z + z\sqrt{z})/(16 - z)$

5. $y = -6$ ($y = -17$ is extraneous)   6. $y = \pm\sqrt{97}$   7. $h = 2$ inches

8. Thinking about it logically, you have never, of course, heard of a
coverup that succeeded because, if it did succeed, then you would have
never heard about it!

9. $-49$

**Imperial**

1. $3y^2\sqrt{5y}$     2. $x = \pm 6$          3. $(3\sqrt{7} + 7)/7$

4. $2x^{34}$

5. $\sqrt[42]{w}$  or  $w^{1/42}$

6. $6\xi$

7. $x = -1$ ($x = -13/4$ is extraneous)

8. $12 - i$

9. $v = \sqrt{0.99}\ c$

# *Lesson Seventeen*

## Venn Diagrams, Disjoint Sets, Union and Intersection of Sets

---

*Life of Fred:*
*Advanced Algebra*
pp. 70–71

1. Here is a Venn diagram where the set B is the set of all Blue Ridge Mountains (which is a mountain range which extends from Virginia to Georgia). On this diagram place a dot (•) where a cheese-and-mushroom omelet would go.

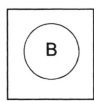

2. Draw a Venn diagram which includes sets C and D. Let C be the set of all coelacanths. *Coelacanth* is pronounced SEE-leh-kanth. It is a truly unique fish. If you ever want to take a half-hour break from advanced algebra, type `coelacanth` into some Internet search engine such as Google. From the beginning of human history, all we knew about coelacanths was from fossils. No one had ever seen a real live coelacanth. We knew was that this four-legged creature liked to crawl around on the ocean bottom. It was the closest animal that was a fish and, if it had been designed just a little differently, would have been an amphibian. Too bad we never saw a live one.

artist rendering of a
coelacanth

Miss Marjorie Courtenay-Latimer worked at a small museum and liked to visit a local fisherman to see if he caught any unusual fish. And, well, you can guess the rest of the story.

And let D be the set of all the dikes in Holland.

3. Solve $\dfrac{x-2}{x+4} = \dfrac{2}{3x+5}$

4. Suppose we have three sets. Call them E, F, and G. Assume that E and F are disjoint and that F and G are disjoint. Must it be true that E and G are disjoint?

5. Suppose we have two sets, H and I. Suppose that neither set is the empty set. (Sometimes this is expressed: Both are nonempty.) Is it possible that the union of H and I could equal the intersection of H and I?

Namely, $H \cup I \overset{?}{=} H \cap I$.

# ANSWERS

1. Since cheese-and-mushroom omelets are not part of the Blue Ridge Mountains, I place the dot outside of the set B.

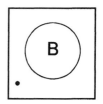

2.

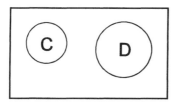

Sets C and D (the sets of all coelacanths and of all dikes in Holland) are disjoint. The circles do not overlap.

3.

$$\frac{x-2}{x+4} = \frac{2}{3x+5}$$

$$3x^2 - 3x - 18 = 0$$

Divide through by 3

$$x^2 - x - 6 = 0$$

$$(x-3)(x+2) = 0$$

$$x = 3, -2$$

4. It might be possible that E and G are disjoint:

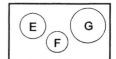

But E and G don't *have* to be disjoint. Suppose that
E = the set of all elephants
F = the set of all flags
G = the set of all animals in Great Britain.
Then the Venn diagram would look like:

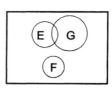

5. It doesn't happen very often, but it can happen. The question is equivalent to asking how can we make the shaded area in [diagram H∪I] equal to the shaded area in [diagram H∩I]

Or, in other words, how can we make every element that is in either set an element of both sets.

The only way this can happen is if the sets are identical. For example, the set H could be those things that are flammable and the set I could be the set of those things that are inflammable. (Both words mean combustible.)

# *Lesson Eighteen*

## Subsets, Using Venn Diagrams in Counting Problems

Do the *Puzzle Questions*.

*Life of Fred:*
*Advanced Algebra*
pp. 72–75

# *Lesson Nineteen*

## Significant Digits

Do the exercises.

*Life of Fred:*
*Advanced Algebra*
pp. 75–81

Odd answers are in
the text, and even
answers are
given here.

2. 2    4. 4    6. 1    8. 3    10. 2    12. 4    14. $9 \times 10^{-2}$
16. $8.8 \times 10^3$    18. $1.037 \times 10^0$    20. $2 \times 10^0$    22. $1.6 \times 10^1$
24. $1.6 \times 10^3$    26. 3980.    28. 7856.    30. 1038.

# *Lesson Twenty*

## Setting Up Exponential Equations

1. Every year you own your truck it loses 12% of its value. Write the exponential equation that would help you find out how many years you would have to own your truck before it lost 80% of its value.

*Life of Fred:*
*Advanced Algebra*
pp. 82–84

2. Your typing speed increases by 5% for each month you practice. Write the exponential equation that would help you find how many months it would take you to double your speed.

1. In one year your truck would be worth 88% of its original value. At the end of the second year it would be worth 88% × 88% of its original value. After x years it would be worth $(0.88)^x$ of its original value.

$$(0.88)^x = 0.2$$

New truck

After one year

2. If your typing speed increases by 5% each month that you practice, your speed after one month would be 1.05 times as much as when you started. After two months it would be $(1.05)^2$ as much as your original speed.

$$(1.05)^x = 2$$

After two years

### *Intermission*

One of the important aspects of mathematics is that it can mirror reality. If my baby gets her first new tooth and then gets another new tooth, then with a quick calculation (1 + 1 = 2), I would expect her to have two teeth.

Increasing my typing speed by 5% each month is an example of exponential growth. The expression $(1.05)^x$ may be a very good model—it may reflect accurately how my speed is increasing *over relatively short periods of time*.

In the first stages of many different things, exponential growth is common. Typists type 5% faster each month. Young companies grow 40% each year. Beginning weight lifters lift 2% more each week.

There is an old expression: *Trees do no grow to the sky*. Translation: Exponential growth may not mirror reality over longer periods of time. Typists do not keep increasing their speed by 5% per month over the years. In four years that typist would be typing more than ten times as fast as before which may not be humanly possible.

Instead, a better mathematical model of what is really happening may be an S-shaped curve: rapid increases at the beginning and a tapering off as time goes by.

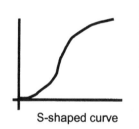

exponential growth

S-shaped curve

# *Lesson Twenty-one*
## Solving Exponential Equations

---

Solve each of these. In case you don't have a scientific calculator with a **log** key on it, I've provided some log values with each problem.

*Life of Fred:*
*Advanced Algebra*
pp. 85–88

1. $(0.93)^x = 0.25$       log 0.93 = –0.0315    log 0.25 = –0.602

2. *You are the Author* ™
Make up a word problem for which $(0.93)^x = 0.25$ is part of the solution.

3. "The Birdie Rule" was only one of many films made by Fred Gauss Productions. Each year this company increased its revenues by 12% over the previous year. How long would it take for the company to triple its current revenues?      log 3 = 0.477    log 1.12 = 0.0492

4. You are trapped on a deserted island without a calculator. You remember that log 4 = 0.602. Find the value of log 16.

5. You are trapped on a deserted island without a calculator. You remember that log 4 = 0.602. Find the value of log 64.

6. You are trapped on a deserted island without a calculator. You remember that log 4 = 0.602. Find the value of log 2.

7. You are trapped on a deserted island without a calculator. You remember that log 4 = 0.602. Find the value of log 0.25.

8. You are trapped on a deserted island without a calculator. You remember that log 4 = 0.602. Find the value of log 1/16.

9. You are trapped on a deserted island without a calculator. You remember that log 4 = 0.602. Find the value of log 1.

And do the *Your Turn to Play*.

1.  $(0.93)^x = 0.25$        $\log 0.93 = -0.0315$    $\log 0.25 = -0.602$

   $\log (0.93)^x = \log 0.25$

   $x \log 0.93 = \log 0.25$

   $x(-0.0315) = -0.602$

   $x = \dfrac{-0.602}{-0.0315}$

   $x \doteq 19.1$

2.  Story #1: Every time the 56-lb. clerk took a sip from her diet water bottle, she drank 7% of what was there. How long before her bottle was only 25% full?

   Story #2: The "cute little froggies" on Fred's hospital gown kept falling off. Each time he looked, there were 93% as many frogs as before. How many times would Fred have to look before there were only a quarter of the number of frogs that he started with?

   Story #3: Each time you solve an exponential equation, it is only 93% as mysterious as it was before. How many exponential equations would you have to solve in order that the whole process lose three-quarters of its mystery?

3.  $(1.12)^x = 3$        $\log 3 = 0.477$    $\log 1.12 = 0.0492$

   $\log (1.12)^x = \log 3$

   $x \log 1.12 = \log 3$

   $x = \dfrac{\log 3}{\log 1.12}$

   $x = \dfrac{0.477}{0.0492} \doteq 9.7$ years

4.  $\log 16 = \log 4^2 = 2 \log 4 = 2 (0.602) = 1.204.$
5.  $\log 64 = \log 4^3 = 3 \log 4 = 3 (0.602) = 1.806$
6.  $\log 2 = \log 4^{\frac{1}{2}} = \frac{1}{2} \log 4 = \frac{1}{2} (0.602) = 0.301$
7.  $\log 0.25 = \log 4^{-1} = - \log 4 = -0.602$
8.  $\log 1/16 = \log 4^{-2} = -2 \log 4 = -2 (0.602) = -1.204$
9.  $\log 1 = \log 4^0 = 0 \log 4 = 0$

   This is one of the likable laws of logs: $\log 1 = 0$

# Lesson Twenty-two
## Product and Quotient Rules for Logs

Life of Fred:
*Advanced Algebra*
pp. 89–94

1. If you start with an investment of $1200 in Waddle Doughnuts, Inc. and it grows at the rate of 6% each year, how long will it take before your investment is worth $5000?

$$\log 1200 = 3.079$$
$$\log 1.06 = 0.025$$
$$\log 5000 = 3.699$$

2. If you know that $\log 8 = 0.90$ and that $\log 10 = 1.0$, find the value of $\log 20$.

3. Complete:

... using the Birdie Rule       $n \log x =$

... using the Product Rule      $\log t + \log w =$

... using both rules      $5 \log y + 6 \log z =$

... using the Product Rule      $\log ac =$

... using the Birdie Rule      $\log (m + n)^4$

4. What, if anything, does $\log (m + n)$ equal? (answers on next page)

And do the *Your Turn to Play*.

# Lesson Twenty-three
## Finding Anti-logs

Do the *Your Turn to Play*.

Life of Fred:
*Advanced Algebra*
pp. 95–99

1.     $1200(1.06)^n = 5000$
$\log\left(1200(1.06)^n\right) = \log 5000$
$\log 1200 + \log (1.06)^n = \log 5000$
$\log 1200 + n \log 1.06 = \log 5000$
$3.079 + n(0.025) = 3.699$
$n \doteq 24.8 \text{ years}$

$\log 1200 = 3.079$
$\log 1.06 = 0.025$
$\log 5000 = 3.699$

2.  $\log 2 = \log 8^{1/3} = (1/3)\log 8 = (1/3)(0.90) = 0.3$
$\log 20 = \log 2(10) = \log 2 + \log 10 = 0.3 + 1.0 = 1.3$

$\log 8 \doteq 0.90$
$\log 10 = 1.0$

3.  . . . using the Birdie Rule          $n \log x = \log x^n$
    . . . using the Product Rule          $\log t + \log w = \log tw$
    . . . using both rules          $5 \log y + 6 \log z = \log y^5 z^6$
    . . . using the Product Rule          $\log ac = \log a + \log c$
    . . . using the Birdie Rule          $\log (m + n)^4 = 4 \log (m + n)$

4. This is the one that fools a lot of beginning log students. (loggers? logarithmers?)  $\log (m + n)$ doesn't equal anything in particular.

It's $\log mn$ that can be expanded (into $\log m + \log n$).

Logger-rhythm

## *Lesson Twenty-four*
### First Definition of Logarithm

Do the *Your Turn to Play.*

*Life of Fred:*
*Advanced Algebra*
p. 100–top of p. 102

## *Lesson Twenty-five*
### Second Definition of Logarithm, The Change-of-Base Rule

Do the *Your Turn to Play.*

*Life of Fred:*
*Advanced Algebra*
bottom of p. 102–p. 104

## *Lesson Twenty-six*
### Third Definition of Logarithm, Logical Implication

1. Change into exponential form: $\log_4 7 = x$.

2. Change into log form: $98^y = 323$

3. Change into log form: $4.1 = z^7$

*Life of Fred:*
*Advanced Algebra*
pp. 105–106

4. Express as a quotient of common logs: $\log_8 65$.

5. Express as a log "What power do you raise 8 to in order to get 65?"

6. Given log 8 = 0.9 and log 65 = 1.81 (for those readers who don't have a scientific calculator). What power do you raise 8 to in order to get 65?

7. Given log 1.07 = 0.029 and log 6 = 0.778 (for those readers who don't have a scientific calculator). What power do you raise 1.07 to in order to get 6?

8. Express in scientific notation: 673.44

9. Suppose we had three surveyors each measure one side of a triangular piece of land. They reported lengths of 7.007, 8.25, and 11 miles. What is the perimeter. Round your answer taking significant digits into account.

10. Same as the previous question except that the reported lengths were 7.007, 8.25, and 11.000 miles.

And do the *Your Turn to Play*.

1. $4^x = 7$

2. $\log_{98} 323 = y$

3. $\log_z 4.1 = 7$

4. $\dfrac{\log_{10} 65}{\log_{10} 8}$   which could also be written as $\dfrac{\log 65}{\log 8}$

5. $\log_8 65$

6. "What power do you raise 8 to in order to get 65?"

   $= \log_8 65$

   $= \dfrac{\log 65}{\log 8}$

   $= \dfrac{1.81}{0.9}$

   $\doteq 2.01$

7. "What power do you raise 1.07 to in order to get 6?"

   $= \log_{1.07} 6$

   $= \dfrac{\log 6}{\log 1.07}$

   $= \dfrac{0.778}{0.029}$

   $\doteq 268$

8. $673.44 = 6.7344 \times 10^2$

9.   $7.007$
     $8.25$
   $+11.$
   ───────
   $26.257$  which rounds to 26 miles.

   ☞ is where we round to

10.  $7.007$
     $8.25$
   $+11.000$
   ───────
   $26.257$  which rounds to 26.26 miles.

   ☞ is where we round to

# *Lesson Twenty-seven*

## End of the Chapter—Review & Testing
## Part One

Do all the problems in the first two cities.

 Electra

San Marcos

> *Life of Fred:*
> *Advanced Algebra*
>
> The Cities starting
> on p. 107

---

# *Lesson Twenty-eight*

## End of the Chapter—Review & Testing
## Part Two

Do all the problems in
the second pair of cities.

Odd answers are in
the text, and even
answers are
given here.

 Ulster

 Wahoo

> *Life of Fred:*
> *Advanced Algebra*
>
> The Cities starting
> on p. 108

## ✹E✹V✹E✹N✹ ✹A✹N✹S✹W✹E✹R✹S✹

Ulster

2. 0

4. $\log 0.001 = \log_{10} 10^{-3} = -3$

6. 3

8. 1.92

Wahoo

2. $\log_2 w$

4. 28 trips    $\log 0.1 = \log_{10} 0.1 = \log_{10} 10^{-1} = -1$

6. 2

8. $y = \dfrac{\log 3}{\log 11}$

# *Lesson Twenty-nine*

### End of the Chapter—Review & Testing
### Part Three

Do all the problems in
the third pair of cities.

*Life of Fred:
Advanced Algebra*

The Cities starting
on p. 109

 Zaleski

 Idaho City

## ★N★S★W★E★R★S

Zaleski

1. $\log_w 3 = 27$      2. 1        3. 1.5

4. 14 years

5. For no values of x, does $\log_x x^2 = 1$.

Using the birdie rule we have $2 \log_x x = 1$ and since

$\log_x x$ is equal to one, we have $2(1) = 1$ which is not possible.

6. 1      7. $2040

Idaho City

1. $8^z = 9$     2. 0     3. 1.4

4. 233 books

5. 1

6. log (−10) asks the question, "To what power do you raise 10 to in order
to get an answer of −10?" But 10 to any power is always positive.

log (−10) is undefined.

7. $x = -0.2772938$

8. $x = \dfrac{\log \sqrt{2}}{\log \pi}$ or $\dfrac{\log 2}{2 \log \pi}$

# Lesson Thirty

## Graphing by Point-Plotting

Do the five exercises on p. 112.
Answers are on the next page.

Life of Fred:
*Advanced Algebra*
pp. 110–112

# Lesson Thirty-one

## Glossary of Graphing Terms

Do the five exercises on p. 114.
Answers are on the next page.

Life of Fred:
*Advanced Algebra*
pp. 113–114

# Lesson Thirty-two

## Slope

1. If you have had some geometry, you'll probably be able to answer: *If a line has a slope equal to one, what angle does the line make with the x axis?*

Life of Fred:
*Advanced Algebra*
pp. 115–117

2. If a line has a slope equal to 3 and the run is equal to 2, what does the rise equal?

3. What is the slope of this line?

4. What is the value of c that would make the slope of this line equal to 5?

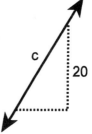

*41*

## ✹A✹N✹S✹W✹E✹R✹S✹ ✹T✹O✹ ✹E✹X✹E✹R✹C✹I✹S✹E✹S✹

1. $y = \cos x$

2. $y = 2x + 4$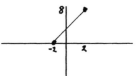

3. $y = 1/x$ 

4. $y = |x|$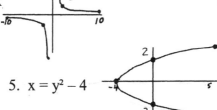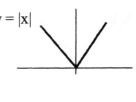

5. $x = y^2 - 4$

6. QII are points $(x, y)$ where $x < 0$ and $y > 0$.

7. If $(a, b)$ where $b > 0$ is not to lie in either QI or QII, it must lie on the positive y axis. Therefore it will be a point such as $(0, 8)$ or $(0, 9792397)$.

8. If $(c, d)$ is in QIII, then both $c$ and $d$ must be negative. Hence $c \times d$ is the product of two negative numbers and is positive.

9. Looking at the graph on the second page of this chapter, we see that $y = \log x$ lies in QI and QIV.

10. The most quadrants that a line (a straight line) can pass through is three. For example:

## ✹A✹N✹S✹W✹E✹R✹S✹

1. When m = 1, then the graph might look like:

2. $\dfrac{\text{rise}}{\text{run}} = m$

$\dfrac{\text{rise}}{2} = 3$     The rise must be 6.

3. By the Pythagorean theorem, $x^2 + 1^2 = 2^2$.     $x^2 = 3$     $x = \sqrt{3}$

(We don't use $x = \pm\sqrt{3}$ since x represents a distance which is never negative.)

 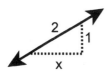 $m = \dfrac{1}{\sqrt{3}}$

4. If m = 5, then x must be 4. By the Pythagorean theorem, c would be $\sqrt{416}$.

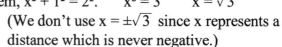

42

# Lesson Thirty-three
## Finding the Slope Given Two Points

Do the *Your Turn to Play.*

> *Life of Fred:*
> *Advanced Algebra*
> pp. 118–121

# Lesson Thirty-four
## Slope-intercept and Double-intercept Forms of the Line

Do the *Your Turn to Play.*

> *Life of Fred:*
> *Advanced Algebra*
> pp. 122–123

# Lesson Thirty-five
## Point-slope and Two-point Forms of the Line

Do the *Your Turn to Play.*

> *Life of Fred:*
> *Advanced Algebra*
> pp. 124–126

# Lesson Thirty-six
## Perpendicular Lines Have Slopes Whose Product Equals −1

Do the *Your Turn to Play*.

> *Life of Fred:*
> *Advanced Algebra*
> pp. 127–130

1. List the five equations from the *Your Turn to Play*.

2. Using the five equations, establish that $m_1 = -1/m_2$.

1. Equation #1:   $m_1 = b/a$
   Equation #2:   $\dfrac{a}{b} = \dfrac{b}{c}$

   Equation #3:   $m_2 = -e/d$
   Equation #4:   $b = e$
   Equation #5:   $c = d$

2. There are many different ways to establish $m_1 = -1/m_2$ using the five equations. It would be rare if two algebraists did it in exactly the same way.

   Here's one approach.

   Cross-multiply the proportion in #2:         $ac = b^2$

   Multiply both sides by $\dfrac{1}{ab}$ :          $\dfrac{ac}{ab} = \dfrac{b^2}{ab}$

   Cancel:                     $\dfrac{c}{b} = \dfrac{b}{a}$

   By the symmetric          $\dfrac{b}{a} = \dfrac{c}{b}$          [What these steps did is invert the
   property of equality                              proportion in equation #2]

   Call $\dfrac{b}{a} = \dfrac{c}{b}$   equation #6.

   Multiply both sides of #3 by $-d$          $-dm_2 = e$
   and by the symmetric property of equality    $e = -dm_2$      (Call this #7.)

   Now here we go. Start with equation #1

   $m_1 = \dfrac{b}{a}$ which equals $\dfrac{c}{b}$ (by #6)

   which equals $\dfrac{d}{b}$ (by #5)

   which equals $\dfrac{d}{e}$ (by #4)

   which equals $\dfrac{d}{-dm_2}$ (by #7)   which equals $-1/m_2$
                                               which was what was wanted.

## *Lesson Thirty-seven*

### End of the Chapter—Review & Testing
### Part One

Do all the problems in the first two cities.

 Cashe Creek

Harlow

> *Life of Fred:*
> *Advanced Algebra*
>
> The Cities starting
> on p. 131

===

## *Lesson Thirty-eight*

### End of the Chapter—Review & Testing
### Part Two

Do all the problems in
the second pair of cities.

> *Life of Fred:*
> *Advanced Algebra*
>
> The Cities starting
> on p. 132

Odd answers are in
the text, and even
answers are
given here.

 Paradise

 Vallejo

### EVEN ANSWERS

Paradise
2. $(y + 1)/(x - 5) = \dfrac{6}{-13}$
4. no and no
6. .........................................

*—Here*

Vallejo
2. $x/5 + y/7 = 1$
4. You can't say anything. It would be positive if the point in QIV were $(1, -1)$ and it would be negative if it were $(100, -1)$. It would be undefined if it were $(5, -7)$. The slope can't be zero.
6. $y = 66$ or $y = 74$

## Lesson Thirty-nine

### End of the Chapter—Review & Testing
### Part Three

Do all the problems in
the third pair of cities.

*Life of Fred:*
*Advanced Algebra*

The Cities starting
on p. 133

 Wakefield

Dairy

⭐A⭐N⭐S⭐W⭐E⭐R⭐S⭐

Wakefield

1. $(y - 15)/(x - 10) = -4/3$
2. 6
3. 8/22
4. the formula that established the distance between $(x_1, y_1)$ and $(x_2, y_2)$
5.

6. c must be zero.


Dairy

1. $(y + 4)/(x - 88) = 2$
2. $\sqrt{13} + \sqrt{17} + 2\sqrt{5}$
3. 10.9/11.3
4. zero
5. $(x/3) + (y/2) = 1$
6. yes, it is unavoidable.

# *Lesson Forty*
## Multiplying Binomials

*Life of Fred:*
*Advanced Algebra*
pp. 134–top two lines of 135

1. $(2y + 7)(3y + 2)$
2. $(6w - 2)(4w - 3)$
3. $(x + 8)(x - 8)$
4. $(11z + 5)^2$
5. What is the slope of $y = 13$?
6. What is the slope of the line that is perpendicular to $y = 5x - 7$?
7. Simplify $\log_{33} 1$.
8. What is the base of $\log_4 6$?
9. Simplify $\log_{12} 4 + \log_{12} 36$.
10. Put in scientific notation: 75.003.
11. Simplify $\sqrt{1000x^7 w}$
12. Give an example of a pure imaginary number.
13. Solve $\sqrt{x^2 - 40} = 3$.

Also do exercises 1 – 10 on p. 135.

1. $(2y + 7)(3y + 2) = 6y^2 + 4y + 21y + 14 = 6y^2 + 25y + 14$
2. $(6w - 2)(4w - 3) = 24w^2 - 26w + 6$
3. $(x + 8)(x - 8) = x^2 - 64$
4. $(11z + 5)^2 = (11z + 5)(11z + 5) = 121z + 110z + 25$
5. It is a horizontal line with a slope of zero.
6. $y = 5x - 7$ has a slope of 5. The line perpendicular to that would have a slope of $-1/5$.
7. $\log_{33} 1$ asks the question $33^? = 1$. The answer is zero.
8. 4 is the base of $\log_4 6$.
9. $\log_{12} 4 + \log_{12} 36 = \log_{12} (4)(36) = \log_{12} 144 = 2$.
10. $75.003 = 7.5003 \times 10^1$
11. $\sqrt{1000x^7w} = \sqrt{100x^6} \sqrt{10xw} = 10x^3 \sqrt{10xw}$
12. Pure imaginary numbers are numbers that look like ri where r is a non-zero real number: $3i$, $17.003987i$, $-0.1i$, $\sqrt{555}\, i$, $\pi i$
13. $\qquad\qquad \sqrt{x^2 - 40} = 3$

Square both sides $\qquad x^2 - 40 = 9$

$$x^2 = 49$$
$$x = \pm 7$$

In solving radical equations, it is necessary to check all answers in the original equations since some of the answers might be extraneous. Putting $x = -7$ into the original equation we get $\sqrt{(-7)^2 - 40} \overset{?}{=} 3$. This is true. Putting $x = 7$ into the original equation also works.

2. $42y^2 - 25y + 3$
4. $x^2 + 4x - 21$
6. $36w^2 - 16$
8. $25x^2 + 30x + 9$
10. $12z^2 + 4.4z + 0.24$

# Lesson Forty-one

## Common Factors

*Life of Fred:*
*Advanced Algebra*
Part B on p. 135

1. Factor $18x + 24$

2. Factor $55x^3 - 44x$

3. Factor $30x^{77}w^{40} + 12x^6w^4 - 60x^7w^9$

4. Multiply out $(4z - 5)(4z + 5)$

5. Multiply out $(6y + 4)^2$

6. Simplify $\log_7 5 + \log_7 (1/5)$

7. Simplify $\log_\pi \pi^3$

8. $\{3, 4, 5\} \cap \{4, 5, 6\}$

9. $\{3, 4, 5\} \cup \{4, 5, 6\}$

10. $(d^5)^6$

11. Rationalize the denominator $\dfrac{\sqrt{7}}{\sqrt{5}}$

12. Rationalize the denominator $\dfrac{4}{5 + \sqrt{6}}$

13. Graph $\dfrac{x}{5} + \dfrac{y}{4} = 1$ [This is in the double-intercept form.]

14. Simplify $81^{1/4}$

15. When "Ready Meddie" broke out her climbing gear, she knew that the amount of rope she would need would be proportional to how high the victim was in the tree. In Prof. Eldwood's *Tree Rescue for Bunnies in Trees,* which was published in 1851, he states that 420 feet of rope is required for a 80-foot-high rescue. What is the constant of proportionality? (This was explained on p. 22.)

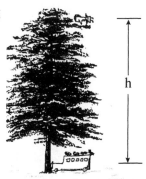

Also do exercises 11 – 20 on p. 135

1. $18x + 24 = 6(3x + 4)$
2. $55x^3 - 44x = 11x(5x^2 - 4)$
3. $30x^{77}w^{40} + 12x^6w^4 - 60x^7w^9 = 6x^6w^4(5x^{71}w^{36} + 2 - 10xw^5)$
4. $(4z - 5)(4z + 5) = 16z^2 - 25$
5. $(6y + 4)^2 = (6y + 4)(6y + 4) = 36y^2 + 48y + 16$
6. $\log_7 5 + \log_7 (1/5) = \log_7 5(1/5) = \log_7 1 = 0$
7. Using the base-eraser approach, $\log_\pi \pi^3 = \log_\pi \pi^3 = \log_\pi \pi^3 = 3$
8. $\{3, 4, 5\} \cap \{4, 5, 6\} = \{4, 5\}$           [∩ = intersection]
9. $\{3, 4, 5\} \cup \{4, 5, 6\} = \{3, 4, 5, 6\}$     [∪ = union]
10. $(d^5)^6 = d^{30}$
11. $\dfrac{\sqrt{7}}{\sqrt{5}} = \dfrac{\sqrt{7}\sqrt{5}}{\sqrt{5}\sqrt{5}} = \dfrac{\sqrt{35}}{5}$
12. $\dfrac{4}{5 + \sqrt{6}} = \dfrac{4(5 - \sqrt{6})}{(5 + \sqrt{6})(5 - \sqrt{6})} = \dfrac{4(5 - \sqrt{6})}{25 - 6} = \dfrac{4(5 - \sqrt{6})}{19}$

13. $\dfrac{x}{5} + \dfrac{y}{4} = 1$

[The double-intercept form was explained on pp. 122–123.]

14. $81^{1/4} = \sqrt[4]{81} = 3$

15. The amount of rope is proportional to how high the victim was in the tree. $A = kh$ where A is the amount of rope and h is the height of the victim in the tree. According to Eldwood, $A = 420$ when $h = 80$.

   Therefore, $420 = k80$.

     $k = 5.25$.

12. $2w^6z^{10}(12w^2 - 5)$
14. $4x^2y^{99}(2x^{64} + 3y)$
16. $2cd(c^{11}d^7 + 11)$
18. $14x^{222}y^5(x^{111}y^{883} - 2 + 3x^{222}y^{83})$
20. $35w^6(w + 2)$

# *Lesson Forty-two*

## Factoring Easy Trinomials, Difference of Squares

*Life of Fred:*
*Advanced Algebra*
Parts C and D on
pp. 135-136

1. Factor $x^2 + 9x + 20$
2. Factor $y^2 + 4y + 4$
3. Factor $w^2 + 8w + 12$
4. Factor $x^2 + 9x + 9$
5. Factor $2y^2 + 10y + 12$
6. Factor $6a^2 - 150$
7. Simplify $144^{1/2}$
8. What is the square of $\sqrt{x(\log_4 y) + 173}$ ?
9. Is 5.55 a rational number?
10. Before "Ready Meddie" began her climb, she told Cheryl that she needed 3.2 cubic yards of water to fill her plastic tree-climbing bottle. "It's important to stay hydrated," she explained.

Convert 3.2 cubic yards into liters.

[There are 202 gallons in a cubic yard. There are 3.785 liters in a gallon.]

I need 3.2 cubic yards of really fresh water

Also do exercises 21–40 on p. 136.

1. $x^2 + 9x + 20 = (x + 5)(x + 4)$
2. $y^2 + 4y + 4 = (y + 2)(y + 2)$    or you could have written $(y + 2)^2$
3. $w^2 + 8w + 12 = (w + 2)(w + 6)$
4. $x^2 + 9x + 8 = (x + 1)(x + 8)$
5. $2y^2 + 10y + 12 = 2(y^2 + 5y + 6) = 2(y + 3)(y + 2)$
6. $6a^2 - 150 = 6(a^2 - 25) = 6(a + 5)(a - 5)$
7. $144^{\frac{1}{2}} = \sqrt{144} = 12$
8. The square of $\sqrt{\xi}$ is $\xi$ for any real number $\xi$.
So the square of $\sqrt{x(\log_4 y) + 173}$ is $x(\log_4 y) + 173$.
9. 5.55 can be written as 555/100. It is a rational number.
10. $\dfrac{3.2 \text{ cubic yards}}{1} \times \dfrac{202 \text{ gallons}}{1 \text{ cubic yard}} \times \dfrac{3.785 \text{ liters}}{1 \text{ gallon}} = 2446.624$ liters

    If you take significant figures into account, you would round this to 2400 liters since there are only two significant digits in 3.2 cubic yards.

22. $(y + 2)(y + 5)$
24. $(w + 7)(w - 3)$.
26. $(x - 4)(x - 7)$ .
28. $5(y^2 + 7y + 25)$
30. $(x^2 + 5)(x^2 + 7)$
32. $(4w + 9y)(4w - 9y)$
34. $250(4x^2 - y^2)$ which factors further as $250(2x + y)(2x - y)$
36. Only a common factor here: $4(25w^2 + 9)$
38. $6(9 - w^{10})$ which factors further as $6(3 + w^5)(3 - w^5)$
40. $(y^4 + 8)(y^4 - 8)$

# *Lesson Forty-three*

## Factoring by Grouping, Factoring Harder Trinomials

---

*Life of Fred:*
*Advanced Algebra*
Parts $E$ and $F$ on
pp. 137–138

1. Factor $x^3 - 9x^2 + 4x - 36$

2. Factor $y^3 + 4y^2 - 9y - 36$

3. Factor $w^2 - w - 30$

4. Factor $2x^2 + 13x + 21$

5. Factor $6x^2 + 23x + 20$

6. Factor $16z^2 - 49$

7. Where does the line $\frac{x}{2} + \frac{y}{6} = 1$ intercept the x axis?

8. What is the distance from $(8, -11)$ to $(3, 7)$?

9. Simplify $\log_{49} 7$.

10. Name the two sets that are disjoint.

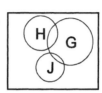

11. Simplify $(4i)^2$

12. Factor $90y^2w^8 - 27y^3w^5 - 30$

13. When the bus driver rammed his bus into the tree, it knocked 40% of the seeds out of the tree. He backed the bus up and hit the tree a second time, and, again, 40% of the seeds that were still in the tree fell to the ground. After hitting the tree a dozen times, what fraction of the seeds would still be in the tree?

14. Express in scientific notation: 0.0003589

Also do exercises 41 – 58 on pp. 137–138.

1. $x^3 - 9x^2 + 4x - 36 = x^2(x - 9) + 4(x - 9) = (x - 9)(x^2 + 4)$
2. $y^3 + 4y^2 - 9y - 36 = y^2(y + 4) - 9(y + 4)$
   $= (y + 4)(y^2 - 9) = (y + 4)(y + 3)(y - 3)$
3. $w^2 - w - 30 = (w - 6)(w + 5)$
4. $2x^2 + 13x + 21 = (2x + 7)(x + 3)$
5. $6x^2 + 23x + 20 = (3x + 4)(2x + 5)$
6. $16z^2 - 49 = (4z + 7)(4z - 7)$
7. $\frac{x}{2} + \frac{y}{6} = 1$ is in the double-intercept form. It intercepts the axis at
at $x = 2$.

8. $(3, 7)$ to $(8, -11)$?  $\sqrt{(3 - 8)^2 + (7 - (-11))^2}\ = \sqrt{25 + 324} = \sqrt{349}$
9. $\log_{49} 7 = 1/2$ since $49^{\frac{1}{2}} = 7$.
10. H and J are disjoint. They have
no elements in common.

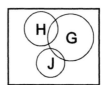

11. $(4i)^2 = (4i)(4i) = 16i^2 = -16$
12. $90y^2w^8 - 27y^3w^5 - 30 = 3(30y^2w^8 - 9y^3w^5 - 10)$
13. After one hit, 60% of the seeds would still be in the tree. After two
hits 60% × 60% of the seeds would still be in the tree. After a dozen hits,
$(0.6)^{12}$ would still be in the tree. Using a calculator, this would be
approximately 0.0021768. (That means he would have shaken 99.78232%
of the seeds out of the tree.)
14. $0.0003589 = 3.589 \times 10^{-4}$

42. $(5x^2 + 2)(2x + 5)$     44. First a common factor $2(6w^3 + 3w^2 + 2w + 1)$
and then by grouping $2(3w^2 + 1)(2w + 1)$
46. After grouping and then a difference of squares $(z + 4)(z - 4)(2z - 5)$
48. $(x + 2)(3x + 7)$        50. $(3x + 2)(x + 1)$
52. $(3x + 1)(x + 6)$        54. $(x + 3)(4x - 3)$
56. This is a "harder trinomial" followed by difference of squares
$(2x + 1)(2x - 1)(2x + 1)(2x - 1)$
58. $(5y - 1)(6y + 1)$

# *Lesson Forty-four*

## Simplifying, Adding, and Subtracting Fractions

Do exercises 59–78

Odd answers are in
the text, and even
answers are
given here.

Life of Fred:
*Advanced Algebra*
Parts G, H and I on
pp. 139–141

### ⭐EVEN⭐ ⭐ANSWERS⭐

60. $(y + 8)/(y - 6)$

62. $(2z - 3)/(z + 7)$

64. $(a - 4)/a^3$

66. $(w - 11)/(3w - 7)$

68. $(3y + 1)/(2y - 1)$

70. $4w/(2w + 3)$

72. $-33/(y + 5)(y - 6)$

74. $(6w + 7)/w^2$

76. $(-y^2 + 38y + 55)/(y - 7)(y + 9)$

78. $(w^2 - w - 6)/w^2$

# *Lesson Forty-five*

## Multiplying and Dividing Fractions

Do exercises 79–84

Odd answers are in
the text, and even
answers are
given here.

Life of Fred:
*Advanced Algebra*
Parts J and K on
pp. 141–142

### ⭐EVEN⭐ ⭐ANSWERS⭐

80. $1/(w - 7)$

82. $(-z - 3)/(3z + 4)$

84. 1

## *Lesson Forty-six*

### Solving Fractional Equations, Solving Quadratic Equations

Do exercises 85–92

Life of Fred:
*Advanced Algebra*
Parts L, M, N, O, and P on
pp. 143–144

Odd answers are in
the text, and even
answers are
given here.

### EVEN ANSWERS

86. $y = 3, -11/3$

88. $w = -3$

90. $x = \pm\sqrt{15}$

92. $y = 5/2, \ 1$

## *Lesson Forty-seven*

### Solving Radical Equations

Do exercises 93–104

Life of Fred:
*Advanced Algebra*
Part Q on p. 144–145

Odd answers are in
the text, and even
answers are
given here.

### EVEN ANSWERS

94. no solution

96. $z = -3$

98. $y = 5$

100. $y = 4$

102. $z = 1, 5$

104. $z = 4$

# *Lesson Forty-eight*

### Systems of Equations, Inconsistent Equations, Dependent Equations

Do the *Your Turn to Play.*

*Life of Fred:*
*Advanced Algebra*
pp. 146–150

# *Lesson Forty-nine*

### Graphing Planes in Three Dimensions

Do the one problem in the *Your Turn to Play* on p. 153.

*Life of Fred:*
*Advanced Algebra*
pp. 151–154

1. Here are three ads. How much does each item cost?

2 bikes
plus
3 cakes
plus
1 dog

All for only
$128

1 bike
plus
2 cakes
plus
1 dog

All for only
$82

3 bikes
plus
5 cakes
plus
1 dog

All for only
$180

Do the one problem in the *Your Turn to Play* on p. 154.

2. Graph the plane $2x + 3y + 5z = 120$

3. $\dfrac{x + 3}{x - 5} - \dfrac{10x + 22}{x^2 - x - 20}$

1.  $2b + 3c + d = 128$     copy     $2b + 3c + d = 128$
    $b + 2c + d = 82$     mult by $-1$     $-b - 2c - d = -82$
    $3b + 5c + d = 180$

    and add them to eliminate the d........ $b + c = 46$

Repeating this with equations one and three

we get.............................................. $-b - 2c = -52$

We now have two equations with two unknowns: $\begin{cases} b + c = 46 \\ -b - 2c = -52 \end{cases}$

Adding these two equations eliminates the b:     $-c = -6$
$c = 6$
Back substituting $c = 6$ into $b + c = 46$, gives us $b = 40$.
Back substituting $c = 6$ and $b = 40$ into $b + 2c + d = 82$, gives us $d = 30$.
The bikes cost \$40; the cakes cost \$6; the dogs cost \$30.

2.

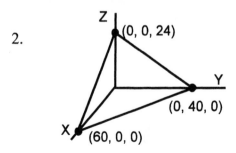

Z
(0, 0, 24)
Y
(0, 40, 0)
X
(60, 0, 0)

3.  $\dfrac{x + 3}{x - 5} - \dfrac{10x + 22}{x^2 - x - 20}$

$= \dfrac{x + 3}{x - 5} - \dfrac{10x + 22}{(x - 5)(x + 4)}$

$= \dfrac{(x + 3)(x + 4)}{(x - 5)(x + 4)} - \dfrac{10x + 22}{(x - 5)(x + 4)}$

$= \dfrac{x^2 + 7x + 12 - (10x + 22)}{(x - 5)(x + 4)} = \dfrac{x^2 - 3x - 10}{(x - 5)(x + 4)} = \dfrac{x + 2}{x + 4}$

# Lesson Fifty
## Cramer's Rule

Do the *Your Turn to Play.*

> Life of Fred:
> *Advanced Algebra*
> pp. 155–156

# Lesson Fifty-one
## 2 × 2 Determinants

1. Evaluate $\begin{vmatrix} 5 & 6 \\ 3 & 4 \end{vmatrix}$

> Life of Fred:
> *Advanced Algebra*
> pp. 157–top half of 159

2. Solve for x using Cramer's rule $\begin{cases} 2x - 3y = 17 \\ 5x + 4y = 8 \end{cases}$

3. $\log 600 - \log 2 - \log 3 = ?$

4. Solve $\dfrac{y+7}{y+9} = \dfrac{2}{3}$

5. Evaluate $\begin{vmatrix} -6 & 3 \\ -3 & 2 \end{vmatrix}$

6. $z^7 z^7 = ?$

7. Fill in the blank: The natural numbers are { ? }.

8. Installing 2500 watt electric outlets on a bus costs $w per outlet. [We encountered a 2500 watt outlet on p. 25 when Fred thought it would be used to power an electric chair.]

Installing AUTOMATIC PILOT buttons on a bus costs $a per button.
Two outlets and one button cost $114.
Five outlets and two buttons cost $236.

Using Cramer's rule, find out how much an AUTOMATIC PILOT button cost.

Do the *Your Turn to Play* on p. 159.

*59*

1. $\begin{vmatrix} 5 & 6 \\ 3 & 4 \end{vmatrix} = 5(4) - 3(6) = 2$

2. $\begin{cases} 2x - 3y = 17 \\ 5x + 4y = 8 \end{cases}$ $\qquad x = \dfrac{D_x}{D} = \dfrac{\begin{vmatrix} 17 & -3 \\ 8 & 4 \end{vmatrix}}{\begin{vmatrix} 2 & -3 \\ 5 & 4 \end{vmatrix}} = \dfrac{68 - (-24)}{8 - (-15)} = 4$

3. $\log 600 - \log 2 - \log 3 = \log \dfrac{600}{2(3)} = \log 100 = 2$

4. $$\dfrac{y + 7}{y + 9} = \dfrac{2}{3}$$

Cross-multiplying $\quad 3(y + 7) = 2(y + 9)$

$$y = -3$$

5. $\begin{vmatrix} -6 & 3 \\ -3 & 2 \end{vmatrix} = (-6)(2) - (-3)(3) = -3$

6. $z^{14}$

7. $\{1, 2, 3, 4, \dots\}$

8. Two outlets and one button cost \$114.
   Five outlets and two buttons cost \$236.

$$\begin{cases} 2w + a = 114 \\ 5w + 2a = 236 \end{cases}$$

$$a = \dfrac{D_a}{D} = \dfrac{\begin{vmatrix} 2 & 114 \\ 5 & 236 \end{vmatrix}}{\begin{vmatrix} 2 & 1 \\ 5 & 2 \end{vmatrix}} = \dfrac{2(236) - 5(114)}{4 - 5} = 98$$

An AUTOMATIC PILOT button cost \$98. That's just the cost of the button. The AUTOMATIC PILOT system, which is activated by the AUTOMATIC PILOT button, costs a little over a zillion dollars. That's why most buses still have human bus drivers. In later editions of this book we may need to delete this paragraph as computers get cheaper.

# Lesson Fifty-two

### 3 × 3 Determinants

Do the *Your Turn to Play.*
Do the *Puzzle Questions.*

*Life of Fred:*
*Advanced Algebra*
pp. 159–164

---

# Lesson Fifty-three

### End of the Chapter—Review & Testing
### Part One

Do all the problems in the first two cities.

*Life of Fred:*
*Advanced Algebra*

The Cities starting
on p. 165

 Sioux Falls

 Cactus

---

# Lesson Fifty-four

### End of the Chapter—Review & Testing
### Part Two

Do all the problems in
the second pair of cities.

*Life of Fred:*
*Advanced Algebra*

The Cities starting
on p. 166

Odd answers are in
the text, and even
answers are
given here.

 Eldorado

 Talmage

---

## EVEN ANSWERS

Eldorado

2. $3x^2 - 2x + 4 = y$    4. neither    6. $x = \dfrac{\begin{vmatrix} 1 & 1 \\ 18 & -2 \end{vmatrix}}{\begin{vmatrix} 1 & 1 \\ 3 & -2 \end{vmatrix}} = \dfrac{-2 - 18}{-2 - 3} = 4$

Talmage

2.  Evaluating by the second column makes things really nice.  The determinant is equal to zero.

4.  hijack = ten years in jail

6.
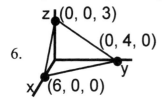

z↓(0, 0, 3)

(0, 4, 0)

x (6, 0, 0)

y

---

## *Lesson Fifty-five*

### End of the Chapter—Review & Testing
### Part Three

Do all the problems in
the third pair of cities.

*Life of Fred:*
*Advanced Algebra*

The Cities starting
on p. 168

 Ulysses

 Yerington

---

## ❶❷❸❹❺❻❼❽

Ulysses

1.  –46        2.  no solution        3.  $2/hour

4.  No it could not be inconsistent since the determinant of the coefficients (D) is negative and therefore not equal to zero.

5.  (–4, 10)        6.  s = 1000, b = 2000, m = 8000

Yerington

1.  The determinant will equal zero.  If you were to expand the determinant by the second row, each of the terms would be 0 times a $22 \times 22$ determinant and hence the final answer would be zero.

2.  The fourth equation lost a minus sign.  It should read $-26x + 10y = 6$.

3. and 4.  white = 40¢; brown = 50¢        5.  102        6.  12

# *Lesson Fifty-six*

## Ellipses

---

Do the *Your Turn to Play* on p. 172.

*Life of Fred:*
*Advanced Algebra*
pp. 171–172

1. What is the length of the semi-major axis of
the ellipse $\frac{x^2}{9} + \frac{y^2}{16} = 1$?

2. Put in standard form $11x^2 + 2y^2 = 22$

3. Put in standard form $3x^2 + 7y^2 = 5$

From arithmetic it may be helpful to recall that $\frac{3}{5}$ can be written $\frac{1}{\frac{5}{3}}$

This is true because $\frac{1}{\frac{5}{3}} = 1 \div \frac{5}{3} = 1 \times \frac{3}{5} = \frac{3}{5}$

4. Looking at the top of a glass, the top edge will look like an ellipse if
you view it from an angle.

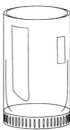

    Let's reverse things. Here is an ellipse.
Is it possible to view it from an angle so that
it looks like a circle?

5. Find the equation of the parabola $ax^2 + bx + c = y$ that passes through
the points $(0, 1)$, $(1, 8)$ and $(2, 19)$.

6. Solve $\dfrac{6}{\sqrt{y^2 - 7}} = 2$

### ⓐⓝⓢⓦⓔⓡⓢ

1. The larger of the two axes is the major axis. For $\dfrac{x^2}{9} + \dfrac{y^2}{16} = 1$ the major axis is in the y direction. The length of the semi-major axis is 4.

2. $$11x^2 + 2y^2 = 22$$

$$\frac{11x^2}{22} + \frac{2y^2}{22} = 1$$

$$\frac{x^2}{2} + \frac{y^2}{11} = 1$$

$$\frac{x^2}{(\sqrt{2})^2} + \frac{y^2}{(\sqrt{11})^2} = 1$$

3. $$3x^2 + 7y^2 = 5$$

$$\frac{3x^2}{5} + \frac{7y^2}{5} = 1$$

$$\frac{x^2}{(\sqrt{5/3})^2} + \frac{y^2}{(\sqrt{5/7})^2} = 1$$

4. If you turn the book sideways and put your eye over here, it will foreshorten the major axis as you look at the ellipse.

5. If (0, 1) is on the parabola $ax^2 + bx + c = y$, then it must satisfy that equation. It must be true that $a(0)^2 + b(0) + c = 1$.

If (1, 8) is on the curve, then $a(1)^2 + b(1) + c = 8$

If (2, 19) is on the curve, then $a(2)^2 + b(2) + c = 19$.

Cleaning these three equations up, we get
$$c = 1$$
$$a + b + c = 8$$
$$4a + 2b + c = 19$$

Solve by Cramer's rule or by the elimination method: $a = 2$, $b = 5$, $c = 1$.
The equation of the parabola is $2x^2 + 5x + 1 = y$.

6. $$\frac{6}{\sqrt{y^2 - 7}} = 2$$

$$6 = 2\sqrt{y^2 - 7}$$
$$3 = \sqrt{y^2 - 7}$$
$$9 = y^2 - 7$$
$$\pm 4 = y \qquad \text{Both answers check in the original problem.}$$

> Checking answers is mandatory in solving radical equations.

# *Lesson Fifty-seven*

## Circles

---

Do the *Your Turn to Play* on p. 177.

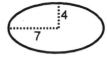

*Life of Fred:*
*Advanced Algebra*
pp. 173–177

1. Write the equation of the circle whose radius is 16.

2. Write the equation of the circle whose diameter is 22.

3. Write the equation of the circle whose radius is $\sqrt{5}$.

4. Write the equation of the ellipse that looks like: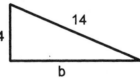

5. Using the birdie rule, $\log_6 7^{35} = ?$

6. Plot $5 + 3i$ on the complex number plane.

7. TRUE-FALSE: The semi-major axis of $\frac{x^2}{21} + \frac{y^2}{387} = 1$ is parallel to the y-axis.

8. Find the length of b:   4

14

b

9. Simplify $\sqrt{180}$ using $\sqrt{xy} = \sqrt{x}\,\sqrt{y}$

10. 5¢ will buy you 0.8 liters of Sluice. How much would a barrel of Sluice cost? [Conversion factors: A barrel = 42 gallons. There are 202 gallons in a cubic yard. One cubic yard contains 764.6 liters.]

11. The sign said that the Sluice machine was 500 yards ahead. How many meters is that? [Conversion factor: 1 meter = 1.094 yards.]

12. What is the equation of the line that passes through (5, 4) and (–6, 3)?

1. $x^2 + y^2 = 256$
2. $x^2 + y^2 = 121$
3. $x^2 + y^2 = 5$
4. $\dfrac{x^2}{49} + \dfrac{y^2}{16} = 1$

5. $\log_6 7^{35} = 35 \log_6 7$

6.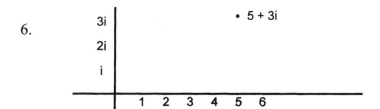

7. Since the larger denominator is under the $y^2$, it is parallel to the y-axis. The statement is true.

8. By the Pythagorean theorem, $4^2 + b^2 = 14^2$
$$b^2 = 180$$
$$b = \sqrt{180} \qquad \text{(Not } \pm\sqrt{180} \text{ since b is a length.)}$$

9. $\sqrt{180} = \sqrt{9}\,\sqrt{20} = 3\sqrt{20} = 3\sqrt{4}\,\sqrt{5} = 6\sqrt{5}$

10. $\dfrac{1 \text{ barrel}}{1} \times \dfrac{42 \text{ gallons}}{1 \text{ barrel}} \times \dfrac{1 \text{ cubic yard}}{202 \text{ gallons}} \times \dfrac{764.6 \text{ liters}}{1 \text{ cubic yard}} \times \dfrac{5\text{¢}}{0.8 \text{ liters}}$

$\doteq 994\text{¢} = \$9.94$

11. $\dfrac{500 \text{ yards}}{1} \times \dfrac{1 \text{ meter}}{1.094 \text{ yards}} \doteq 457 \text{ meters}$

12. The two-point form of the line is $\dfrac{y - y_1}{x - x_1} = \dfrac{y_2 - y_1}{x_2 - x_1}$
where $(x_1, y_1)$ and $(x_2, y_2)$ are points.
The line that passes through $(5, 4)$ and $(-6, 3)$ is $\dfrac{y - 4}{x - 5} = \dfrac{3 - 4}{-6 - 5}$
This simplifies to $x - 11y + 39 = 0$.

Suppose our goat was at the point (x, y). We want the sum of the distances from the goat to the two stakes (the two foci) to be a constant.

*Life of Fred:*
*Advanced Algebra*
pp. 178–top half of 181

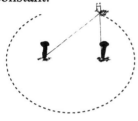

If our goat is at point P and the foci are points F and F', then we want PF + PF' = some constant. [In geometry, the distance from point P to point F was written as PF.]

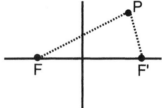

What we want to show is that if our goat moves so that the sum of its distances to the two stakes remains constant, then it *must* move in the path of an ellipse.

To make things easier, let's say that PF + PF' equals $\sqrt{8}$. (I chose $\sqrt{8}$ because that makes things as simple as possible. If I had chosen 1 instead, we would have to deal with a lot of fractions. In order to make this perfectly general, we could have said PF + PF' equals a constant c, but as Scarlett O'Hara said, "We'll think about that another day.") Place the foci at (–1, 0) and (1, 0).

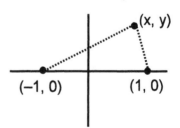

Now for the *hardest problem of this study guide*: Starting from the fact that the distance from (x, y) to (–1, 0) plus the distance from (x, y) to (1, 0) is equal to $\sqrt{8}$, show that $\frac{x^2}{2} + \frac{y^2}{1} = 1$.

[You'll need to use the distance formula from p. 121 and the technique for solving equations with two radicals found at the bottom of p. 145. It will take about a dozen lines of algebra.]

## ✹A✹N✹S✹W✹E✹R✹

If you could solve this before you looked at the answer given below, you are in the top 1% of all algebra students.

By the distance formula, the distance from (x, y) to (1, 0) is $\sqrt{(x-1)^2 + (y-0)^2}$ .

By the distance formula, the distance from (x, y) to (–1, 0) is $\sqrt{(x+1)^2 + (y-0)^2}$ .

Since we are told that the sum of these two distances is equal to $\sqrt{8}$, our equation is

$$\sqrt{(x-1)^2 + (y-0)^2} + \sqrt{(x+1)^2 + (y-0)^2} = \sqrt{8}$$

$$\sqrt{(x-1)^2 + y^2} + \sqrt{(x+1)^2 + y^2} = \sqrt{8}$$

The instructions from p. 145: *If you have two radicals in an equation, isolate one of the radicals and square both sides.* We transpose . . .

$$\sqrt{(x+1)^2 + y^2} = \sqrt{8} - \sqrt{(x-1)^2 + y^2}$$

and square . . .

$$(x+1)^2 + y^2 = 8 - 2\sqrt{8}\sqrt{(x-1)^2 + y^2} + (x-1)^2 + y^2$$

Multiply out . . .

$$x^2 + 2x + 1 + y^2 = 8 - 2\sqrt{8}\sqrt{(x-1)^2 + y^2} + x^2 - 2x + 1 + y^2$$

Simplify . . .

$$2\sqrt{8}\sqrt{(x-1)^2 + y^2} = 8 - 4x$$
$$\sqrt{8}\sqrt{(x-1)^2 + y^2} = 4 - 2x$$

Square both sides again . . .

$$8((x-1)^2 + y^2) = 16 - 16x + 4x^2$$
$$8(x^2 - 2x + 1 + y^2) = 16 - 16x + 4x^2$$
$$8x^2 - 16x + 8 + 8y^2 = 16 - 16x + 4x^2$$
$$4x^2 + 8y^2 = 8$$
$$\frac{x^2}{2} + \frac{y^2}{1} = 1$$

## *Lesson Fifty-nine*
### Reflective Property of Ellipses

Do the *Your Turn to Play.*

*Life of Fred:*
*Advanced Algebra*
Bottom half of p. 181–184

## *Lesson Sixty*
### Parabolas

Do the *Your Turn to Play.*

*Life of Fred:*
*Advanced Algebra*
pp. 185–186

1. Looking at the graphs on p. 186, predict what the graph of $y = 100x^2$ will look like.

2. If we were to graph $y = 100x^2$ by point plotting, we would name some values for x and then find the corresponding values for y. Fill in the rest of this chart:

| Points for the equation $y = 100x^2$ | | | | | | | |
|---|---|---|---|---|---|---|---|
| x | −3 | −2 | −1 | 0 | 1 | 2 | 3 |
| y | | 400 | | | | | |

3. Here is your ordinary parabola $y = x^2$. Its vertex is located at the origin.

Here is that same parabola which has moved so that its vertex is at (6, 3).
$$y - 3 = (x - 6)^2$$

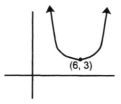

(6, 3)

What would be the equation if we wanted the vertex to be at (55, 8)?

4. $\dfrac{9x^2 - 4}{15x + 10} \div \dfrac{6x^2 - 13x + 6}{2x^2 + 5x - 12}$

1. We could expect $y = 100x^2$ to look like:

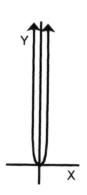

2.

| Points for the equation $y = 100x^2$ | | | | | | | |
|---|---|---|---|---|---|---|---|
| x | −3 | −2 | −1 | 0 | 1 | 2 | 3 |
| y | 900 | 400 | 100 | 0 | 100 | 400 | 900 |

3. $y - 8 = (x - 55)^2$

4. $\dfrac{9x^2 - 4}{15x + 10} \div \dfrac{6x^2 - 13x + 6}{2x^2 + 5x - 12}$

$= \dfrac{9x^2 - 4}{15x + 10} \times \dfrac{2x^2 + 5x - 12}{6x^2 - 13x + 6}$

$= \dfrac{(3x + 2)(3x - 2)(2x - 3)(x + 4)}{5(3x + 2)(3x - 2)(2x - 3)}$

$= \dfrac{x + 4}{5}$

# Lesson Sixty-one

## Hyperbolas

---

Do the *Your Turn to Play*.

*Life of Fred:*
*Advanced Algebra*
pp. 187–189

1. One branch of a hyperbola does not have
the same shape as a parabola. In problem 1 of the *Your Turn to Play*, you
sketched $\frac{x^2}{49} - \frac{y^2}{4} = 1$.

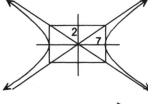

If you were to look at that graph
from ten feet away, it would look like:
It starts to look like a greater than sign, >,
trying to kiss a less than sign, <. (There's
a lot of romance in mathematics if you
look carefully.)

If you sketched $x = 2y^2$ and then

tried to draw in asymptotes, you would
be frustrated. No matter what line you
chose, the parabola would bend away from that line.

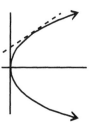

Your question: The curve $y = \frac{1}{x}$ has two asymptotes.
By sketching the curve, find them.

1.  Using point plotting, we first make a table and get some points on the curve:

| Points for the equation y = 1/x | | | | | | | | | |
|---|---|---|---|---|---|---|---|---|---|
| **x** | 1 | 2 | 3 | 0.1 | 0.01 | −1 | −2 | −10 | −0.1 | −0.01 |
| **y** | 1 | 1/2 | 1/3 | 10 | 100 | −1 | −1/2 | −0.1 | −10 | −100 |

Then we plot those points:

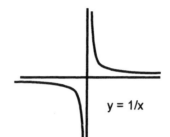

The asymptotes are the
x-axis and the y-axis.

This curve is a hyperbola that has been rotated 45°.

---

## *Lesson Sixty-two*
### Graphing Inequalities, Definition of a Conic Section

---

Do the *Your Turn to Play.*

*Life of Fred:*
*Advanced Algebra*
pp. 190–196

# *Lesson Sixty-three*

## End of the Chapter—Review & Testing
### Part One

Do all the problems in the first two cities.

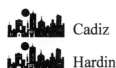 Cadiz

 Hardin

> *Life of Fred:*
> *Advanced Algebra*
>
> The Cities starting
> on p. 197

# *Lesson Sixty-four*

## End of the Chapter—Review & Testing
### Part Two

Do all the problems in
the second pair of cities.

> *Life of Fred:*
> *Advanced Algebra*
>
> The Cities starting
> on p. 198

Odd answers are in
the text, and even
answers are
given here.

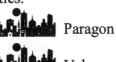 Paragon

 Vale

## EVEN ANSWERS

### Paragon

2.

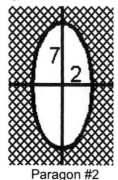

Paragon #2

4. If the semi-minor axis is vertical, then the semi-major axis must be horizontal. Since by definition, the major axis is larger than the minor axis, then $a^2$ must be larger than $b^2$.

6. $(-1, 7)$ and $(9, 7)$

### Vale

2. $(-7, -5 \pm \sqrt{33}\,)$   4a. $2a$   4b. $a$

4c. $c = \sqrt{a^2 - b^2}$   6. $(3, -5)$

# *Lesson Sixty-five*

## End of the Chapter—Review & Testing
### Part Three

Do all the problems in
the third pair of cities.

> *Life of Fred:*
> *Advanced Algebra*
>
> The Cities starting
> on p. 199

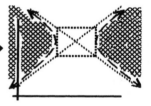 Walbridge

Dakota City

★A★N★S★W★E★R★S★

Walbridge
1. $(x - 6)^2 + (y - 11)^2 = 25$
2. $(3, 6 \pm \sqrt{13})$
3.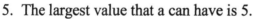
4. $(x - (-4))^2 + (y - 7)^2 = (\sqrt{6})^2$
5. The largest value that a can have is 5.

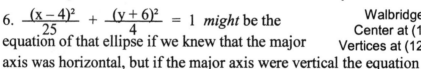

Walbridge # 3
Center at (12, 10)
Vertices at (12 ± 7, 10)

6. $\dfrac{(x - 4)^2}{25} + \dfrac{(y + 6)^2}{4} = 1$ *might* be the

equation of that ellipse if we knew that the major
axis was horizontal, but if the major axis were vertical the equation

would be $\dfrac{(x - 4)^2}{4} + \dfrac{(y + 6)^2}{25} = 1$.

Dakota City
1.

Center at (5, 6)
Radius = 2

2. $(55 - \sqrt{7}, 66)$ and $(55 + \sqrt{7}, 66)$
3. $(x - (-6))^2/1 + (y - 2)^2/4 = 1$,
    center at $(-6, 2)$
4. $k \geq 8$
5. $b^2 > a^2$
6. $3/2$

# *Lesson Sixty-six*
## Domain, Codomain, Definition of Function

*Life of Fred:*
*Advanced Algebra*
pp. 200–203

1.  Suppose our domain was the set of these three famous modern mathematicians: {Garrett Birkhoff,* Paul Halmos,** Peter J. Hilton***}. (This list could easily be extended to include scores of other famous modern mathematicians. I've just picked these three almost at random.)

Suppose our codomain is {Euclid****}. Create a function with these two sets.

2.  Let the domain = {John Horton Conway, Martin Gardner, Benoit Mandelbrot} and let the codomain = {Solomon Lefschetz, Stanislaw M. Ulam}. (These are five more modern mathematicians.) Create a function with these two sets.

3.  Harder question: How many different possible functions could you have created to answer problem 2?

4.  If the domain has 35 elements and the codomain has 67 elements, how many different functions are possible?

---

✱ Garrett Birkhoff has written new mathematics in many different areas: classical analysis, the four color problem, number theory, lattice theory—approximately 190 articles plus at least five books.

✱✱ Paul Halmos has written ten books and about a hundred papers. He entered the University of Illinois when he was fifteen.

✱✱✱ When Peter J. Hilton was ten years old, he was run over by a car—a Rolls Royce! He spent a long time in the hospital recovering. To pass the time he played with math problems, writing them on his cast. Hilton discovered that he really liked math. Over the years he wrote almost 300 papers in fields such as algebraic topology and homological algebra. One of his earliest books was *Introduction to Homotopy Theory*.

✱✱✱✱ Euclid lived about 300 B. C. He wrote a geometry book entitled *The Elements*. More than 2000 editions of *The Elements* have been published. Only one book, the Bible, has had more editions. The second most "edition-full" book in the world and it's a geometry book!

**ANSWERS**

1. There is only one possible function when the domain = {Garrett Birkhoff, Paul Halmos, Peter J. Hilton} and the codomain = {Euclid}.
Here it is:      Birkhoff → Euclid
                 Halmos → Euclid
                 Hilton → Euclid


2. There are several possible functions.
One possibility:   Conway → Lefschetz
                   Gardner → Ulam
                   Mandelbrot → Ulam


Another:           Conway → Lefschetz
                   Gardner → Lefschetz
                   Mandelbrot → Lefschetz


Another:           Conway → Ulam
                   Gardner → Ulam
                   Mandelbrot → Lefschetz


(Here is some English. Looking at this last example, we say that Conway is *mapped* to Ulam by this function. We say that Lefschetz is *the image* of Mandelbrot under this function.)

3. There are two possible images for Conway (either Ulam or Lefschetz). There are two possible images for Gardner. There are two possible images for Mandelbrot.

---

### Intermission

If there are two possibilities—waffles or pancakes—on the menu and there are three possible syrups—strawberry, cherry and maple—then there are 6 (= 2 × 3) possible breakfasts.

If you own three shirts, four pair of pants and six hats, there are 3 × 4 × 6 = 72 possible outfits you could wear.

---

So with two possible images for Conway, two for Gardner and two for Mandelbrot, there are $8 = 2 \times 2 \times 2$ possible functions.

4. $67^{35}$

# Lesson Sixty-seven

## Is This a Function?

---

Do the *Your Turn to Play* on p. 206.

Life of Fred:
Advanced Algebra
pp. 204–206

Let the domain be the set of the seven Dust
Bunnies = {Meddie, Rita, Fredrika, Alpha, Beta,
Gamma, Delta}. Let the codomain be these three cities
in Oklahoma = {Buffalo, Muskogee, Elmer}.

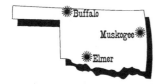

Which of the following rules would be a function?

1. Assign each Dust Bunny to her favorite city in Oklahoma.

2. Assign each Dust Bunny to Elmer.

3. Assign each Dust Bunny to the city in the codomain that is closest
to the city that she was born in.

4. Assign each Dust Bunny to the element of the codomain that has
the same number of letters as her name.

5. Assign each Dust Bunny to the city in the codomain that she knows
the most about.

6. Can the codomain have fewer elements in it than the domain?

7. Can the codomain have more elements in it than the domain?

8. The **range** of a function is the set of images in the codomain. The
range is the elements in the codomain that are "hit" by elements in the
domain. Draw a Venn diagram of the sets range and codomain.

# ❶❷❸❹❺❻❼❽

1. Beta's favorite city is El Reno, Oklahoma. But that's not in the codomain. El Reno became her favorite city in Oklahoma after she heard about its famous fried onion burgers. These fried onion burgers have been cooked daily since the early 1900s. On the first Saturday in May they have a Fried Onion Burger Day festival. They cook the Big Burger—the world's largest fried onion hamburger—and invite festival goers to a free bite. The burger weighs more than 850 pounds.

   In any event, each element of the domain must be assigned to one element in the codomain. El Reno is not in the codomain. This would not be a function.

2. This would look like:
   Meddie → Elmer
   Rita → Elmer
   Fredrika → Elmer
   Alpha → Elmer
   Beta → Elmer
   Gamma → Elmer
   Delta → Elmer

   This fits the definition of function since each element in the domain has exactly one image in the codomain.

3. Meddie, Rita, Fredrika, Alpha, Beta, and Gamma were born in Tennessee and under this rule would be assigned to Muskogee. Delta was born in Oregon and would be assigned to Buffalo under this rule. Since each element in the domain is assigned to exactly one element of the codomain, this is a function.

4. Where would Rita be assigned? No element of the codomain has four letters. It's not a function.

5. Alpha has never heard about Buffalo, Muskogee or Elmer. Where would she be assigned? It's not a function.

6. Look at problem 2 above.

7. On p. 203 of the text, we assigned each of the seven Dust Bunnies to her favorite color. The domain had seven elements = {Meddie, Rita, Fredrika, Alpha, Beta, Gamma, Delta}. The codomain was the set of all colors. There are thousands of different colors. This was a function where the codomain had many more elements than the domain.

8. In general, it would look like:  In the case that every element of the codomain was an image of some element of the domain it would look like:

Do the *Your Turn to Play* on p. 208

```
Life of Fred:
Advanced Algebra
pp. 207–209
```

1. Let function f be given by the rule f(x) = 2x + 7. The domain and codomain are the real numbers. What would f(3) equal?

2. Let the domain and codomain be the set of all married people. Let function s be given by the rule: s(x) = the person x is married to. (I chose *s* as the name of this function to remind me of *spouse*.)

What would s(Mrs. Mittens) equal? [We're assuming a monogamous world.]

3. Let s be the function defined in the previous problem. Let h be the function from the set of all people to the set of real numbers. Define h by the rule: assign to each person his or her current weight in pounds. (I chose *h* as the name of this function to remind me of *heavy*.) Suppose that we know that Mrs. Mittens weighs 136 pounds and Mr. Mittens weighs 190 pounds.

What would h(Mrs. Mittens) equal?

What would h(s(Mrs. Mittens)) equal?

What would s(Mr. Mittens) equal?

What would h(s(Mr. Mittens)) equal?

4. Continuing the previous problem, what would be your response if someone asked you to evaluate s(h(Mr. Mittens))?

5. Continuing the previous problems, what would s(s(Mr. Mittens)) equal?

6. What is the equation of the circle whose center is (5, 88) and whose radius is 12?

7. Graph $\dfrac{x^2}{16} + \dfrac{y^2}{81} \leq 1$

8. Graph $\dfrac{x^2}{16} + \dfrac{y^2}{81} < 1$

9. If g:{all cows}→{★, ✱}, what is the domain of this function?

**❄A❄N❄S❄W❄E❄R❄S❄**

1. f(3) = 2(3) + 7 = 13.
2. Mr. Mittens
3. h(Mrs. Mittens) = 136
   h(s(Mrs. Mittens)) = h(Mr. Mittens) = 190
   s(Mr. Mittens) = Mrs. Mittens
   h(s(Mr. Mittens)) = h(Mrs. Mittens) = 136

---

### Intermission

Back in beginning algebra days, when you wanted to simplify something like 5(7 + 3(x + 4)), the easiest way was *to start on the inside and work your way out.*
You start with the underlined: 5(7 + 3(x + 4)).
This became 5(7 + 3x + 12).

One of my students once called this the Cancer Rule. I asked her what she meant. She smiled and said, "It's like cancer: It starts on the inside and works its way out." The rest of the class groaned.

When you want to evaluate h(s(Mrs. Mittens)), the place to start is with the underlined: h( s(Mrs. Mittens) ).

---

4. To evaluate s(h(Mr. Mittens)), you would start with the underlined: s( h(Mr. Mittens) ). This would equal s(190). Now we get into trouble. The expression s(190) asks what is the person that 190 is married to. That's silly. The domain of s is the set of all married people, and the number 190 is *not* in that set. s(190) is not defined. It is like asking the question: *What blandorf could quave a friggle into three pieces?*
5. s(s(Mr. Mittens)) = s(Mrs. Mittens) = Mr. Mittens
6. (x – 5)² + (y – 88)² = 144

7.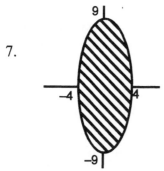

8.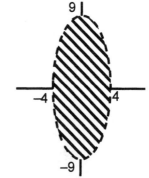

9. In the notation f:A→B, the set A is the domain and the set B is the codomain. In the case of g: {all cows}→{☆, ✳}, the domain = {all cows}.

# *Lesson Sixty-nine*

## One-to-one Functions, Inverse Functions

Do the *Your Turn to Play*.

*Life of Fred:*
*Advanced Algebra*
pp. 210–213

1. We know that

Meddie → apple pie
Rita → strawberry pie
Fredrika → blueberry pie
Alpha → blackberry pie
Beta → orange marmalade pie
Gamma → cherry pie
Delta → lime pie

is a 1-1 function. How could you change this so that it wouldn't be 1-1?

2. Suppose you had seven daughters. (It happens!) And suppose there are nine toys in the room where your daughters are playing. The domain will be your daughters and the codomain will be the nine toys. Define function f by the rule that assigns to each daughter the toy that she wants to play with *right now*. Explain why you hope that f is 1-1.

3. Let the domain be the set of all humans who were alive in the year 1940 and the codomain be the set of all mothers. Define g by the rule: g(x) = the mother of x. Is g a 1-1 function?

4. Continuing the previous problem. Is g a function?

5. George Mittens decided to eat a whole pie. Fred couldn't believe that was going to happen. He said to George, "It would take me a month and a half to eat that much. There are 50 square inches on the top of that pie."

What is the radius of that (circular) pie? (For this problem, approximate π by 3.14 and round your answer to the nearest tenth of an inch.

6. George's waistline was increasing at the rate of 7% per year. How long would it take for him to double his waistline? (If you don't have a calculator with a **log** key on it, leave your answer in terms of common logs.)

1. There are many possible ways to do this. For example, if you changed *Rita → strawberry pie* to *Rita → apple pie*, then it would no longer be one-to-one since Meddie is also mapped to apple pie. Two elements of the domain would be mapped to the same element in the codomain.

2. If f is 1-1 that means that no two elements of the domain (no two daughters) will want to be playing with the same toy *right now*. That means that there will be peace and quiet.

3. Minnie Marx had five sons: Chico, Harpo, Groucho, Gummo, and Zeppo. Since g(Chico) = Minnie and g(Harpo) = Minnie, g is *not* a one-to-one function.

4. Yes, since everyone who was alive in 1940 has exactly one (biological) mother, g is a function. (You are a very sharp reader if you can figure out why I put "who was alive in 1940" in the description of the domain.)

5. The area of a circle is $A = \pi r^2$ where r is the radius.

$$50 = 3.14r^2$$
$$\frac{50}{3.14} = r^2$$
$$\sqrt{\frac{50}{3.14}} = r \qquad \text{(We don't use } \pm \text{, since r is a distance.)}$$
$$r \doteq 4.0"$$

6.
$$1.07^x = 2$$

Take the log of both sides
$$\log 1.07^x = \log 2$$

Birdie rule
$$x \log 1.07 = \log 2$$

$$x = \frac{\log 2}{\log 1.07}$$

And using the **log** key:
$$x \doteq \frac{0.30103}{0.02938} \doteq 10.2 \text{ years}$$

# *Lesson Seventy*

## Guess the Function

---

Do the *Your Turn to Play* on p. 216

```
-----------------------------
|      Life of Fred:        |
|    Advanced Algebra       |
|      pp. 214–215          |
-----------------------------
```

1. Here is a function that maps ordered pairs to the integers. Can you guess the function?

$(3, 4) \rightarrow 7$    $(5, 3) \rightarrow 8$    $(2, 2) \rightarrow 4$    $(8, 5) \rightarrow 13$    $(5, 5) \rightarrow 10$

$(11, 0) \rightarrow 11$  $(4, 2) \rightarrow 6$    $(1, 88) \rightarrow 89$  $(3, 0) \rightarrow 3$    $(10, 0) \rightarrow 10$

---

### *Intermission*

In beginning algebra we might have started with an equation such as $x - 4 = 13$. We added 4 to each side and found that $x = 17$.

We gave as a reason something like, "If equals are added to equals, then the results are equal." That seemed true (and it is) and everyone nodded. If it had been a classroom situation, the students would have looked like:

☺☺☺☺☺☺☺☺☺☺☺☺☺☺☺☺☺☺☺☺☺☺☺☺☺☺☺☺☺☺☺☺☺☺☺☺☺☺☺☺
☺☺☺☺☺☺☺☺☺☺☺☺☺☺☺☺☺☺☺☺☺☺☺☺☺☺☺☺☺☺☺☺☺☺☺☺☺☺☺☺
☺☺☺☺☺☺☺☺☺☺☺☺☺☺☺☺☺☺☺☺☺☺☺☺☺☺☺☺☺☺☺☺☺☺☺☺☺☺☺☺

And almost everyone was happy except maybe ✦.

If you asked ☹ why he was unhappy, he might have asked, "Why is it true? Why, if you add 4 to things that are equal, do you get equal answers?"

You might have said, "Wait till we get to advanced algebra. Then we can explain it."

Well . . . this *is* advanced algebra. It's time to do the explaining.

Explanation: "Adding 4" is a function. Its domain and codomain are all possible numbers. By definition of *function,* each element in the domain is mapped to exactly one element in the codomain. If we start with the number 13, it is mapped to 17. It's always mapped to 17—even in February. When we are given that $x - 4 = 13$, we know that $x - 4$ is another name for 13. So if 13 is mapped to 17 under this function, then $x - 4$ must also be mapped to 17.

And if there were such a thing as the Explanation Function, it might map ☹ → ☺.

---

2. Guess the Function is a fun game to play. You can play it at parties or on the bus. It doesn't have to involve numbers. Just make sure that each element in the domain has exactly one image in the codomain. You give examples and your opponent tries to guess the function. The only caution is that your function rule should be *guessable.* Can you think of a function that no one could guess?

# ⓐⓝⓢⓦⓔⓡⓢ

1. This is one of the first functions you learned in elementary school. It's called addition.

2. First, let me give you some examples of functions that would be *guessable*—where your opponent would have a decent chance of guessing what rule you had in mind.

First example of a function you might use in the Guess the Function game:

    Keanu Reeves → Keanu Reeves

    Chuck Norris → Carlos Ray

    Judy Garland → Frances Gumm

    Marlon Brando → Marlon Brando

    Martin Sheen → Ramon Estevez

    Walter Matthau → Walter Matuschanskavasky

    Tom Cruise → Thomas Cruise Mapother IV

    Cary Grant → Archibald Leach

    Groucho Marx → Julius Henry Marx

[answer: their stage names are mapped to their real names.]

Second example of a function you might use:

    Rob Roy → pineapple pizza

    Harpo Marx → pineapple pizza

    Richard Nixon → squash

    Billy Graham → pineapple pizza

    William J. Clinton → squash

    Florence Nightingale → pineapple pizza

[answer: map to pineapple pizza unless they were a United States president, in which case, you map them to squash.]

Here's my example of a function that is not guessable. I don't think anyone in the whole world would be able to guess this, even if I gave them a thousand examples: (The domain will be last names of people.)

Here's my rule: If their last name appears in the 1948 edition of the New York phone book, select the first occurrence of the name. If it doesn't appear, use the name Wallis. Note the page number. Call this number y. Using Prof. Eldwood's 1855 *Alphabetical List of French Cheeses,* select the $y^{th}$ name on that list. (If there are more than y names in the list, subtract 83 from y as many times as necessary.) Select the penultimate letter in that cheese name. For example, Gauss appears on p. 287 of the 1948 edition of the N.Y. phone book. The $287^{th}$ cheese in Eldwood's listing is Safran du Quercy. Its penultimate letter is *c*. Therefore, Gauss → c.

## *Lesson Seventy-one*
### The Story of the Big Motel—Adding One to Infinity

Do the *Your Turn to Play.*

> *Life of Fred:*
> *Advanced Algebra*
> pp. 216–220

## *Lesson Seventy-two*
### Onto Functions, One-to-one Correspondences

Do the *Your Turn to Play.*

> *Life of Fred:*
> *Advanced Algebra*
> pp. 221–228

## *Lesson Seventy-three*
### Functions as Ordered Pairs, Relations, the Identity Function

Do the *Your Turn to Play.*

> *Life of Fred:*
> *Advanced Algebra*
> pp. 229–231

## *Lesson Seventy-four*
### End of the Chapter—Review & Testing
### Part One

Do all the problems in the first two cities.

 Cainsville

Eleanor

> *Life of Fred:*
> *Advanced Algebra*
> The Cities starting
> on p. 232

## Lesson Seventy-five

### End of the Chapter—Review & Testing
### Part Two

Do all the problems in
the second pair of cities.

Odd answers are in
the text, and even
answers are
given here.

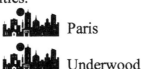

Paris

Underwood

*Life of Fred:*
*Advanced Algebra*

The Cities starting
on p. 233

---

## ● E V E N ● ● A N S W E R S ●

Paris

2. It isn't a function since 8 is mapped to both 2 and 9.

4. Yes since no two real numbers under the rule $f(x) = 1/x$ are mapped to the same image.

6. There is only one possible function from a domain with nine elements to a codomain with one element. It would be the function that maps all the elements of the domain to that one element of the codomain.


Underwood

2. This is not possible. There are too many elements in the range. If we had said that the codomain is {1, 2, 3, . . . , 99, 100}, then this could be done.

4. $h^{-1}$ is not a function since h is not 1-1. I have two daughters. Under function h, each of them is assigned to me, but $h^{-1}$(me) would have two images and hence not be a function.

6. No. Fred is not the biological father of anyone.

*86*

# *Lesson Seventy-six*

### End of the Chapter—Review & Testing
### Part Three

Do all the problems in
the third pair of cities.

| *Life of Fred:* |
| --- |
| *Advanced Algebra* |
| The Cities starting |
| on p. 234 |

 Yoakum

 Zane Hills

## A N S W E R S

Yoakum

1. This would be a function. That fact that it is not onto the codomain doesn't matter.

2. It is a function since no two ordered pairs have the same first coordinate. Its domain is {228, 0, 3}.

3. Only B) is a function. In A) there are two points that have the same x coordinate. Suppose these two points were (5, 7) and (5, 8). That would mean that 5 is mapped to 7 and also to 8. This violates the definition of function: *Every element of the first set is mapped to exactly one element of the second set.* In C) there are many pairs of points that are vertically aligned.

4. There are many possible answers. For example, you could map all the red apples to A, the green apples to B and the others to C. A second example: map all the apples weighing less than 3 ounces to A; those weighing more than 3 ounces but less than 4 ounces to B and the rest to C. A third example: all the apples with less than two seeds to A; those with two seeds to B and the rest to C.

5. $f^{-1}$ B → A is not a function for two different reasons. First, f is not 1-1, so there are elements of B which would have more than one image under $f^{-1}$. Second, f is not onto B, so there are elements of B which would not have any image under $f^{-1}$.

6. Yes. The identity function is always 1-1 since no two different elements in the domain could have the same image.

Zane Hills

1. The range is {3, 4, 5}. The fact that the function is onto means that the range is the same as the codomain. That fact that the function is not the identity function is irrelevant.

2. Example A is a function that is onto but not 1-1.

Example B is not a function since ♣ is not assigned an image in the codomain.

Example C is a function that is neither 1-1 nor onto.

3. It is not 1-1. A counterexample: (5, 4) and (8, 7) are both mapped to 1.

4. Some high school students in Zane Hills have never seen even one Jeanette MacDonald movie and those students would have no image under g. Hence g would not be a function.

5. $g(h(7)) = g(3) = 27$

6. $h(g(7)) = h(343) = 339$

7. One example of an unguessable function would be: Let the domain be the natural numbers and the codomain be {0, 1, 2, 3, . . . , 8, 9} and the rule would be take the $\log_{7.48}$ of that number[*] and then give the 18th digit after the decimal as the image.

A second example: let the domain be the set of all people now living. Let the rule be to take the sum of the number of cars that that person owns plus the number of times they have turned on a television set in the last 22 hours minus six times the number of hot dogs they've eaten in the last month plus 5.3976067.

A third example: let the domain be the set of all 1997 Cadillacs and let the rule be the product of the number of miles on the odometer and the weight of the car (to the nearest pound) and the number of tape cassettes in the car plus the fourth digit of the vehicle identification number. If the resulting answer is evenly divisible by 13, then add 2 to your answer.

---

[*] To find $\log_{7.48} 5$, you could use the change-of-base rule: $\log_b a = \dfrac{\log_c a}{\log_c b}$

$$\log_{7.48} 5 = \frac{\log_{10} 5}{\log_{10} 7.48}$$

Do exercises 1–8.

(Answers to the even exercises are on the next page.)

Life of Fred:
*Advanced Algebra*
p. 236

1. Is it possible to have a function from all movie stars which is onto the natural numbers = {1, 2, 3, 4, . . .}?

2. Where is the center of a circle which has a diameter with endpoints (4, 15) and (4, 25)?

3. Graph $\dfrac{(x-4)^2}{9} + \dfrac{(y-6)^2}{16} \leq 1$

Consider this set of ordered pairs:

{(Meddie, 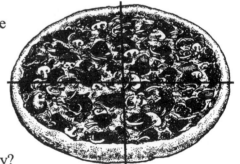), (Rita, 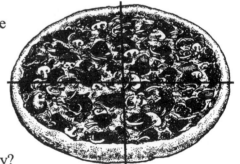 ), (Meddie, 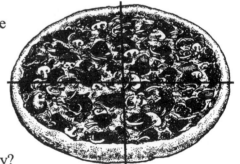 )}

    4. Is this a function?

    5. Is this a relation?

If you put a set of axes through the house-special super-ultra-deluxe-combination pizza, you would note that the one piece of anchovy is located at (–4, –5) and the one piece of lamb was at (3, –2). (We are measuring everything in inches.)

    6. In what quadrant is the anchovy?

    7. In what quadrant is the lamb?

    8. What is the distance between these two toppings?

    9. What is the slope of the line that joins these two toppings?

    10. What is the abscissa of the anchovy?

    11. What is the ordinate of the lamb?

2. 4    4. 36    6. 4    8. 9


★A★N★S★W★E★R★S★

1. A function whose domain is all movie stars would have at most a finite number of images. Jeanette MacDonald might be mapped to 398792. Danny Kaye might be mapped to 89955. But the function would not be *onto* all of the natural numbers since not every natural number would be image of some star.

2. (4, 20)

center is at (4, 6)

3. $\dfrac{(x-4)^2}{9} + \dfrac{(y-6)^2}{16} \leq 1$

(4, 10)

(1, 6)    (7, 6)

(4, 2)

4. It's not a function since Meddie is mapped to two different images.

5. It is a relation since every set of ordered pairs is (by definition) a relation.

6. The point (−4, −5) is located in QIII.

7. The point (3, −2) is located in QIV.

8. The distance between $(x_1, y_1)$ and $(x_2, y_2)$ is given by the formula
$$d = \sqrt{(x_2 - x_1)^2 + (y_2 - y_1)^2}$$
The distance between (−4, −5) and (3, −2) is $\sqrt{(3-(-4))^2 + (-2-(-5))^2}$
$= \sqrt{58}$ which is approximately 7.6 inches. Some people say that the anchovy and the lamb shouldn't be this close to each other. Some say that the anchovy should be in the other room.

9. $m = \dfrac{y_2 - y_1}{x_2 - x_1} = \dfrac{-2 - (-5)}{3 - (-4)} = \dfrac{3}{7}$

10. The abscissa of (−4, −5) is the first coordinate which is −4.

11. The ordinate of (3, −2) is the second coordinate which is −2.

## *Lesson Seventy-eight*
### Long Division of Polynomials

Do exercises 9–12.

*Life of Fred:*
*Advanced Algebra*
pp. 237–239

Odd answers are in
the text, and even
answers are
given here.

10.  $3x^2 - 2x + 9 + 39/(4x - 2)$

12.  $x^3 - 3x^2 + 9x + 2 - 6/(x + 3)$

## *Lesson Seventy-nine*
### Partial Fractions

Do the *Your Turn to Play.*

*Life of Fred:*
*Advanced Algebra*
pp. 240–246

## *Lesson Eighty*
### Proofs by Math Induction

Do the *Your Turn to Play.*

*Life of Fred:*
*Advanced Algebra*
pp. 247–254

1. By math induction prove that
   $1 + 3 + 5 + 7 + \ldots + (2n - 1) = n^2$ is true for every natural number n.

2. Prove $1^3 + 2^3 + 3^3 + \ldots + n^3 = \dfrac{n^2(n + 1)^2}{4}$ for every natural number n.

1.  There are two steps in a math induction proof:

       Step 1: Show that it is true for $n = 1$.

       Step 2: Show that if it is true for $n = k$, it is also true for $n = k + 1$.

Step 1: To show $1 + 3 + 5 + 7 + \ldots + (2n - 1) = n^2$ is true for $n = 1$.

For $n = 1$, $1 + 3 + 5 + 7 + \ldots + (2n - 1) = n^2$ becomes $1 = 1^2$. That's true.

Step 2: We assume $n = k$ is true. I.e., $1 + 3 + 5 + 7 + \ldots + (2k - 1) = k^2$.

We want to establish that $n = k + 1$ is true.

We want to show that $1 + 3 + 5 + 7 + \ldots + (2(k + 1) - 1) = (k + 1)^2$ is true.

Multiplying out the left and right sides, we want to show that

    $1 + 3 + 5 + 7 + \ldots + (2k + 2 - 1) = k^2 + 2k + 1$

    $1 + 3 + 5 + 7 + \ldots + (2k + 1) \quad\ = k^2 + 2k + 1$

What's the odd number just before $2k + 1$?

Since odd numbers are two apart from each other, it would be $(2k + 1) - 2$, which is $2k - 1$. Inserting that number into the above list we get:

    $1 + 3 + 5 + 7 + \ldots + (2k - 1) + (2k + 1) = k^2 + 2k + 1$

We want to show that the above equation is true.

I'll underline the first part of that equation:

    $\underline{1 + 3 + 5 + 7 + \ldots + (2k - 1)} + (2k + 1) = k^2 + 2k + 1$

By the given part of Step 2, we know that the underlined part equals $k^2$.

$$\underline{\quad\quad\quad k^2 \quad\quad\quad} + (2k + 1) = k^2 + 2k + 1 \text{ which is}$$

obviously true. The induction step (Step 2) is done.

2. For $n = 1$, $1^3 + 2^3 + 3^3 + \ldots + n^3 = \dfrac{n^2(n + 1)^2}{4}$ becomes

$$1^3 = \dfrac{1(1 + 1)^2}{4} \text{ which is true.}$$

We now assume $n = k$ is true: $1^3 + 2^3 + 3^3 + \ldots + k^3 = \dfrac{k^2(k + 1)^2}{4}$

We will try to show $n = k + 1$ is true.

We will try to show $1^3 + 2^3 + 3^3 + \ldots + (k + 1)^3 = \dfrac{(k + 1)^2((k + 1) + 1)^2}{4}$

which is $1^3 + 2^3 + 3^3 + \ldots + k^3 + (k + 1)^3 = \dfrac{(k + 1)^2((k + 1) + 1)^2}{4}$

which (by the induction hypothesis "$n = k$ is true") is

$$\dfrac{k^2(k + 1)^2}{4} + (k + 1)^3 = \dfrac{(k + 1)^2((k + 1) + 1)^2}{4}$$

Factor $(k + 1)^2$ out: $(k + 1)^2 \left[\dfrac{k^2}{4} + (k + 1)\right] = \dfrac{(k + 1)^2(k + 2)^2}{4}$

$$(k + 1)^2 \ \dfrac{k^2 + 4k + 4}{4} \ = \dfrac{(k + 1)^2 \ (k^2 + 4k + 4)}{4}$$

# *Lesson Eighty-one*

## Plotting the Constraints for Linear Programming, Big Numbers

---

Do the *Your Turn to Play*.

*Life of Fred:*
*Advanced Algebra*
pp. 255–258

1. Plot on a single graph the four inequalities:

$x + 3y \leq 6, \ 2x + y \leq 7, \ x \geq 0, \ y \geq 0$

2. Okay. You want big numbers. Bigger than Skewes' number which is

$10^{10^{10^{34}}}$ which equals $10^{10^{10000000000000000000000000000000000}}$ which, in turn, is a lot bigger than a googolplex = $10^{10^{100}}$.

In 1976 Donald Knuth invented the **arrow notation**. For natural numbers a and b, he defined $a \uparrow b = a^b$. That's easy. Then he wrote $a \uparrow \uparrow b$

and defined that as $a^{a^{a^{.^{.^{.^{a}}}}}}$ where there are b copies of a in that **power tower**.

More recently, John Conway generalized Knuth's arrow notation. He invented the **chained arrow notation**. For any natural numbers:

(i) $a \rightarrow b$ means $a^b$.

(ii) $a \rightarrow b \rightarrow c \rightarrow \ldots \rightarrow m \rightarrow 1 \quad = \quad a \rightarrow b \rightarrow c \rightarrow \ldots \rightarrow m$
  (You can always chop a "$\rightarrow 1$" off the right side.)

(iii) $a \rightarrow b \rightarrow \ldots \rightarrow m \rightarrow 1 \rightarrow \ldots \rightarrow z \quad = \quad a \rightarrow b \rightarrow \ldots \rightarrow m$
  (A "1" deletes everything to its right up to a right parenthesis.)

(iv) $a \rightarrow b \rightarrow c \quad = \quad a \rightarrow (a \rightarrow b - 1 \rightarrow c) \rightarrow c - 1$
  (For chains of length 3.) E.g., $8 \rightarrow 3 \rightarrow 5 = 8 \rightarrow (8 \rightarrow 2 \rightarrow 5) \rightarrow 4$

(v) $a \rightarrow b \rightarrow \ldots \rightarrow p \rightarrow q \rightarrow r \quad =$
  $a \rightarrow b \rightarrow \ldots \rightarrow p \rightarrow (a \rightarrow b \rightarrow c \rightarrow \ldots \rightarrow p \rightarrow q - 1 \rightarrow r) \rightarrow r - 1$
  (For chains of length 4 or more.)

Some notes:

♪#1: $4 \rightarrow 7$ is the same as $4 \uparrow 7$ which is $4^7$.

♪#2: $5 \rightarrow 9 \rightarrow 2$ is the same as $5 \uparrow \uparrow 9$ which is $5^{5^{5^{5^{5^{5^{5^{5^{5}}}}}}}} = 5^{5^{5^{5^{5^{5^{3125}}}}}} = 5^{5^{5^{5^{(1.9 \times 10^{2184})}}}}$

♪#3: This stuff is pretty tough to understand.

♪#4: Let's see how the rules work:  Start with $2 \rightarrow 3 \rightarrow 3$

$= 2 \rightarrow (2 \rightarrow 2 \rightarrow 3) \rightarrow 2$ (by rule iv)

$= 2 \rightarrow (2 \rightarrow (2 \rightarrow 1 \rightarrow 3) \rightarrow 2) \rightarrow 2$ (by rule iv)

$= 2 \rightarrow (2 \rightarrow \ \ \ 2 \ \ \ \rightarrow 2) \rightarrow 2$ (by rule iii)

> Always work with the innermost set of parentheses.

(continued next page)

| | | | |
|---|---|---|---|
| I'll put in gray the stuff that I'm just copying and not working on. | $= 2 \rightarrow (2 \rightarrow (2 \rightarrow 1 \rightarrow 2) \rightarrow 1) \rightarrow 2$ | (by rule iv) |
| | $= 2 \rightarrow (2 \rightarrow \quad 2 \quad \rightarrow 1) \rightarrow 2$ | (by rule iii) |
| | $= 2 \rightarrow (2 \rightarrow \quad 2 \quad ) \rightarrow 2$ | (by rule ii) |
| | $= 2 \rightarrow \quad 4 \quad \rightarrow 2$ | (by rule i) |
| | $= 2 \rightarrow (2 \rightarrow 3 \rightarrow 2) \rightarrow 1$ | (by rule iv) |

Your question: Do the next two lines of this.

♪#5: Note that all of the lines of ♪#4 are all chains of length 3 or less.

♪#6: Chains of length 4 are mind-numbingly hard. It's been said that no one (and no computer) has any idea what $5 \rightarrow 5 \rightarrow 5 \rightarrow 5$ equals. I think it's near infinity. If you wrote out the answer it would easily have more digits than there are atoms in the observable universe. Which means, of course, that it would be physically impossible to write it out since it takes more than one atom to write a digit. But that was even true of a plain old googol = $10^{100}$. The number $5 \rightarrow 5 \rightarrow 5 \rightarrow 5$ is soooooo large that there is no way to express it using exponents or even power towers. No one—not even Fred—has a real feeling for how big $5 \rightarrow 5 \rightarrow 5 \rightarrow 5$ is.

3. Do the first step in evaluating $7 \rightarrow 3 \rightarrow 9 \rightarrow 5$.

## ⓐⓝⓢⓦⓔⓡⓢ

1.

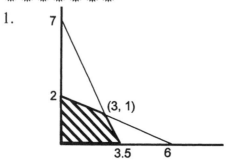

2. We start with $2 \rightarrow (2 \rightarrow 3 \rightarrow 2) \rightarrow 1$

$\quad\quad = 2 \rightarrow (2 \rightarrow 3 \rightarrow 2)$ (by rule ii)

$\quad\quad = 2 \rightarrow (2 \rightarrow (2 \rightarrow 2 \rightarrow 2) \rightarrow 1)$ (by rule iv)

3. $7 \rightarrow 3 \rightarrow 9 \rightarrow 5$

$\quad = 7 \rightarrow 3 \rightarrow (7 \rightarrow 3 \rightarrow 8 \rightarrow 5) \rightarrow 4$ (by rule v)

# Lesson Eighty-two

## The Second Half of Linear Programming

1. Suppose Rita received four big ones for each radio interview and only three big ones for each verse that she wrote. What would be the profit function?

*Life of Fred:*
*Advanced Algebra*
pp. 259–261

2. Continuing the previous problem, suppose that the constraints were the same as those given in the book. Then the graph would still look like →

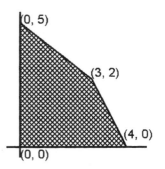

How many interviews and how many verses should Rita do in order to maximize her profit?

3. Suppose you are making pies and cakes. Each pie requires a cup of flour and each cake requires two cups of flour. You only have 12 cups of flour in the house. Let x = number of pies you make. Let y = number of cakes you make. Write the three inequalities and graph them on a single graph. (Hint: one of the inequalities is x ≥ 0—you can't make less than zero pies.)

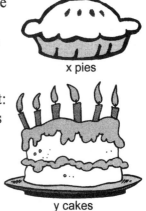

x pies

y cakes

4. Invent a function whose domain is the natural numbers and whose range is the negative integers.

5. Why wouldn't f(x) = –7 (which maps every natural number to –7) be a correct answer to the previous problem?

6. Find a different answer to problem 4.

7. Invent a function whose domain is the natural numbers and whose codomain is the negative integers which is not 1-1.

8. Invent a function whose domain is the natural numbers and whose codomain is the negative integers which is 1-1 but not onto the negative integers.

**ANSWERS**

1.  $f(x, y) = 4x + 3y$

2.  We insert each of the vertices of the graph into the profit function:

$f(0, 0) = 4(0) + 3(0) = 0$      not much profit here

$f(0, 5) = 4(0) + 3(5) = 15$      fifteen billion bucks if she does no interviews and writes 5 verses

$f(3, 2) = 4(3) + 3(2) = 18$      eighteen billion bucks if she does 3 interviews and writes 2 verses

$f(4, 0) = 4(4) + 3(0) = 16$      sixteen billion bucks if she does 4 interviews

Rita should do 3 interviews and write 2 verses.

3.

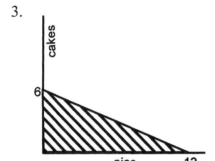

4.  Most people give as their answer $f(x) = -x$ which maps $1 \rightarrow -1$, $2 \rightarrow -2$, $3 \rightarrow -3$, $4 \rightarrow -4$, etc.

5.  If the problem were to find a function whose domain is the natural numbers and whose *codomain* is the negative integers, $f(x) = -7$ would have worked. Since it said that the *range* is to be the negative integers, then every negative integer had to be "hit," not just $-7$.

6.  This is harder. Here's one possibility: $1 \rightarrow -2$, $2 \rightarrow -1$, and then map all the others as before: $3 \rightarrow -3$, $4 \rightarrow -4$, $5 \rightarrow -5$, $6 \rightarrow -6$, etc.

7.  The function $f(x) = -7$ will do the job.

Another possibility is to map the numbers from 1 to 9 all to $-1$, the numbers from 10 to 19 all to $-2$, the numbers from 20 to 29 all to $-3$, etc. This function is not 1-1, but it is onto the negative integers.

8.  There are many possibilities. For example, map $1 \rightarrow -11$, $2 \rightarrow -12$, $3 \rightarrow -13$, $4 \rightarrow -14$, etc. This could be written as $f(x) = -x - 10$.

## *Lesson Eighty-three*
### End of the Chapter—Review & Testing
### Part One

Do all the problems in the first two cities.

 Calder

Emery

*Life of Fred:*
*Advanced Algebra*

The Cities starting
on p. 262

## *Lesson Eighty-four*
### End of the Chapter—Review & Testing
### Part Two

Do all the problems in
the second pair of cities.

Odd answers are in
the text, and even
answers are
given here.

 Garrison

San Mateo

*Life of Fred:*
*Advanced Algebra*

The Cities starting
on p. 264

## ✸EVEN✸ ✸ANSWERS✸

Garrison

2. ($\text{ll},0$)   4. For n = 1 we have $2 = 3^1 - 1$, which is true. Assume n = k is true. To show n = k + 1 which is $2 + 6 + \ldots + 2(3^{k+1-1}) = 3^{k+1} - 1$.
By the assumption the left side reduces to $3^k - 1 + 2(3^{k+1-1})$. This equals $(3^k) - 1 + 2(3^k)$. Combining the $(3^k)$ with the $2(3^k)$ we get $3(3^k) - 1$ which is $3^{k+1} - 1$.

San Mateo

2. (1, 2)

4. You'll employ Pat for 12 hours and Chris for 18 hours.

# *Lesson Eighty-five*

## End of the Chapter—Review & Testing
## Part Three

---

Do all the problems in
the third pair of cities.

> *Life of Fred:*
> *Advanced Algebra*
>
> The Cities starting
> on p. 265

Zalma

Dale

---

## ⓐⓝⓢⓦⓔⓡⓢ

Zalma

1. $\dfrac{7}{x-2} + \dfrac{5}{x+3}$

2. $(0, 5)$

3. You should read 20 chapters of each.

4. For $n = 1$, the statement becomes $5 \stackrel{?}{=} 2(1)^2 + 3(1)$ which is true.

We assume the $n = k$ statement is true: $5 + 9 + \ldots + 4k + 1 = 2k^2 + 3k$.

We will show the $n = k + 1$ statement is true:

$$5 + 9 + \ldots + 4(k + 1) + 1 \stackrel{?}{=} 2(k + 1)^2 + 3(k + 1).$$

Using the assumption we replace the first $k$ terms by $2k^2 + 3k$:

$$2k^2 + 3k + 4(k + 1) + 1 \stackrel{?}{=} 2(k + 1)^2 + 3(k + 1)$$

Multiplying out each side: $2k^2 + 3k + 4k + 4 + 1 \stackrel{?}{=} 2k^2 + 4k + 2 + 3k + 3$
which is true.

Dale

1. $(4, 0)$   2. $\dfrac{x+3}{x^2+2} + \dfrac{5}{2x-1}$   3. No shirts and five cartloads of pants.

4. The $n = 1$ statement ($1 \stackrel{?}{=} \dfrac{3(1)^2 - 1}{2}$ ) is true.

Assume the $n = k$ statement is true: $1 + 4 + \ldots + 3k - 2 = \dfrac{3k^2 - k}{2}$

The $n = k + 1$ statement: $1 + 4 + \ldots + 3(k + 1) - 2 \stackrel{?}{=} \dfrac{3(k+1)^2 - (k+1)}{2}$

Using the assumption: $\dfrac{3k^2 - k}{2} + 3(k + 1) - 2 \stackrel{?}{=} \dfrac{3(k+1)^2 - (k+1)}{2}$

Working with each side: $\dfrac{3k^2 - k + 6(k+1) - 4}{2} \stackrel{?}{=} \dfrac{3k^2 + 6k + 3 - k - 1}{2}$

# *Lesson Eight-six*

## Arithmetic Progressions

Do the *Your Turn to Play*.

> *Life of Fred:*
> *Advanced Algebra*
> pp. 267–269

1. Fill in one word: A sequence is a bunch of numbers separated by

_____.

2. Fill in one word: An series is a bunch of numbers separated by
_____ signs.

3. Fill in the blank in this arithmetic sequence: 44, 56, __?__.

4. Why is this *not* an arithmetic sequence? $35 + 40 + 45 + 50 + 55$

5. Fill in one word: An arithmetic progression means the same as an
arithmetic _____?

6. If the first and third terms of an arithmetic sequence are 7 and 17, what
is the second term?

7. If the first two terms of an arithmetic sequence are 8 and 88 and $n = 30$,
find $a$, $d$, $l$ and $s$.

8. If the first and third terms of an arithmetic sequence are 7 and 16, what
is the second term?

9. If the first and fiftieth terms of an arithmetic sequence are 7 and 700,
what is the common difference $d$?

10. Graph $\dfrac{(x-5)^2}{16} + \dfrac{(y-3)^2}{9} = 1$

11. Graph $\dfrac{(x-5)^2}{16} - \dfrac{(y-3)^2}{9} = 1$

# ✸A✸N✸S✸W✸E✸R✸S✸

1. A **sequence** is a bunch of numbers separated by commas. For example, 3, 987, 998766, 45972349629, 927349029232964335555.

2. A series is a bunch of numbers separated by plus signs. If someone tells you that $1 - 3 + 5 - 7 + 9 - 11 + \ldots$ is a series, that's okay. All you need to do is write it as $1 + (-3) + 5 + (-7) + 9 + (-11) + \ldots$.

3. 44, 56, 68. The first two numbers are 12 apart from each other so the next number is $56 + 12$.

4. It's not separated by commas. $35 + 40 + 45 + 50 + 55$ is an arithmetic series.

5. An arithmetic progression means the same as an arithmetic series.

6. 7, 12, 17, 22, . . . is an arithmetic sequence where the terms are 5 apart.

7. $a$ is the first term which is 8. Since the second term is 88, the common difference $d$ is 80. The last term, $l$, is $= a + (n - 1)d$ which is $8 + (29)80 = 2328$. The sum of the series is $s = \dfrac{n}{2}(a + l) = \dfrac{30}{2}(8 + 2328) = 35{,}040$.

8. 7, 11.5, 16, 20.5, . . . is an arithmetic sequence where the terms are 4.5 apart.

9. We know that $a = 7$ and that $l = 700$ and that $n = 50$.
The formula $l = a + (n - 1)d$ becomes $700 = 7 + (49)d$. After a little algebra, $d = \dfrac{693}{49}$

10. $\dfrac{(x - 5)^2}{16} + \dfrac{(y - 3)^2}{9} = 1$

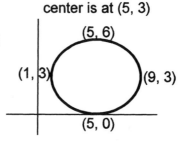

center is at (5, 3)

(5, 6)

(1, 3)          (9, 3)

(5, 0)

11. $\dfrac{(x - 5)^2}{16} - \dfrac{(y - 3)^2}{9} = 1$

(5, 6)

(1, 3)          (9, 3)

(5, 0)

*100*

# *Lesson Eighty-seven*
## Adding and Multiplying Matrices

Do the *Your Turn to Play*.

*Life of Fred:*
*Advanced Algebra*
pp. 270–274

1. If a matrix has three rows and five columns, it is called a "three by five" matrix. Sometimes this is written 3 × 5. These are the **dimensions** of the matrix. Fill in one word: To add together two 3 × 5 matrices, you add corresponding _____ together.

2. The *Fred's Famous For Fine Fountain Pens* stores stocked various colors of fountain pen ink. In the Cuba store there were 5 bottles of red ink, 13 of blue ink, 7 of green, and 2 of violet. In the Lampasas store there were 3 reds, 14 blues, 6 greens, and 2 violets. Write a matrix to summarize this inventory. Let the stores be the rows.

3. The inks originally sold for
      red $3, blue $4, green $4, violet $6.
   After the tariff was put in place, the prices became
      red $4, blue $8, green $6, violet $9.

   Write a matrix to summarize these selling prices. Let the prices before and after the tariff be the columns.

4. If you were to multiply the matrices of problems 2 and 3 together, what would the answer represent?

5. Do it. (Multiply those matrices together.)

6. If someone bought all the inks at the Lampasas store after the tariff was put in place, how much more would he have had to pay than if he bought it before the tariff?

## Intermission

A tariff is a tax on goods imported into the country. It raises prices for the consumer. Take refined sugar for example. In the United States we pay 250% more than the world price because we do not have free trade.

[ Source: http://www.sugar.ca/fact_intNatlPrt.htm ]

7. If you make a bunch of fudge and it would have cost a dollar on the world market for the sugar in the fudge, how much would it cost in the United States?

1. To add together two 3 × 5 matrices, you add corresponding entries together.

2.
$$\begin{pmatrix} 5 & 13 & 7 & 2 \\ 3 & 14 & 6 & 2 \end{pmatrix}$$

3.
$$\begin{pmatrix} 3 & 4 \\ 4 & 8 \\ 4 & 6 \\ 6 & 9 \end{pmatrix}$$

4. The four entries in the product matrix would be the total worth of the inventory of each store before and after the tariff.

5.
$$\begin{pmatrix} 5\cdot3 + 13\cdot4 + 7\cdot4 + 2\cdot6 & 5\cdot4 + 13\cdot8 + 7\cdot6 + 2\cdot9 \\ 3\cdot3 + 14\cdot4 + 6\cdot4 + 2\cdot6 & 3\cdot4 + 14\cdot8 + 6\cdot6 + 2\cdot9 \end{pmatrix}$$

$$= \begin{pmatrix} 107 & 184 \\ 101 & 178 \end{pmatrix}$$

6. After the tariff, the Lampasas store's inventory of inks was $178. Before the tariff, it was $107. He would have to have paid $71 more (= 178 − 107).

7. If the price were 100% higher in the United States, it would have cost $2. If it were 200% higher, it would have cost $3. Since it's 250% higher, it would cost $3.50.

# *Lesson Eighty-eight*
## Geometric Sequences

---

Do the *Your Turn to Play*.

```
Life of Fred:
Advanced Algebra
pp. 275–276
```

1. Classify each of these as an arithmetic series, a geometric series or neither:

    A) 6, 8, 10, 12, . . .
    B) 10, 100, 1000, 10000, . . .
    C) 4, –8, 16, – 32, 64, . . .
    D) 1.5, 2.6, 3.7, 4.8, . . .
    E) 1, 1.1, 1.11, 1.111, 1.1111, . . .
    F) 1, 0.1, 0.01, 0.001, . . .
    G) 1, 1.01, 1.02, 1.03, 1.04, . . .

2. What is the $35^{th}$ term in the geometric sequence 7, 7/5, 7/25, . . . ?

3. If you owned a calculator with a **log** key on it, you could find out approximately how much $7(1/5)^{34}$ is equal to. (Without a scientific calculator you would have to punch in $7 \div 5 \div 5 \div 5 \div 5 \div 5 \div 5 \div 5 \div 5 \div 5 \div 5 \div 5 \div 5 \div 5 \div 5 \div 5 \div 5 \div 5 \div 5 \div 5 \div 5 \div 5 \div 5 \div 5 \div 5 \div 5 \div 5 \div 5 \div 5 \div 5 \div 5 \div 5 \div 5 \div 5.$)

    Using the product rule, we know that $\log 7(1/5)^{34}$ equals what?

4. Continuing the previous problem, using the birdie rule (also known as the exponent rule) we know that $\log (1/5)^{34}$ equals what?

5. Finishing up the previous two problems, how would you compute $7(1/5)^{34}$?

6. What is the $107^{th}$ term of 4, 6, 9, 13.5, . . . ? If you have a scientific calculator, use the **log** and $10^x$ keys to approximate your answer.

---

### *Intermission*

Using the log and $10^x$ keys gives you only an *approximate* answer. When you tapped in log 4 on your calculator you got 0. 60206. The actual value of log 4 is more like 0.6020599913279623. . . . It's an unending non-repeating decimal like $\pi$.

---

1.  A) 6, 8, 10, 12, . . . is an arithmetic sequence with d = 2.
    B) 10, 100, 1000, 10000, . . . is a geometric sequence with r = 10.
    C) 4, –8, 16, – 32, 64, . . . is a geometric sequence with r = –2.
    D) 1.5, 2.6, 3.7, 4.8, . . . is an arithmetic sequence with d = 1.1.
    E) 1, 1.1, 1.11, 1.111, 1.1111, . . . is neither.
    F) 1, 0.1, 0.01, 0.001, . . . is a geometric sequence with r = 0.1.
    G) 1, 1.01, 1.02, 1.03, . . . is an arithmetic sequence with d = 0.01.
2.  The geometric sequence 7, 7/5, 7/25, . . . could be written as
$$7, 7(1/5), 7(1/5)^2, 7(1/5)^3, 7(1/5)^4. \ldots$$
The first term, $a = 7$. The common ratio $r = 1/5$. We want the 35th term which means that $n = 35$. The formula, which we found in the fifth problem of the *Your Turn to Play*, is $l = ar^{n-1}$.
$$l = 7(1/5)^{34}$$
3.  $\log 7(1/5)^{34} = \log 7 + \log (1/5)^{34}$.
4.  $\log (1/5)^{34} = 34 \log (1/5)$.
5.  Here are all the steps: First you take the log of $7(1/5)^{34}$.
Then $\log 7(1/5)^{34} = \log 7 + \log (1/5)^{34} = \log 7 + 34 \log (1/5)$.
Then using the log key this would equal $0.845098 + 34(-0.69897)$ which equals $-22.919882$.

We took the log of the original expression, $7(1/5)^{34}$, and found it is equal to $-22.919882$. We take the antilog and we'll have our answer. The antilog (also known as the $10^x$ key) is usually found on calculators right above the log key. The antilog of $-22.919882$ on my calculator reads: 1.2026 $-23$ which means $1.2026 \times 10^{-23}$ or 0.000000000000000000000012026.

6.  4, 6, 9, 13.5, . . . is a geometric progression with $a = 4$, $r = 1.5$ and we want the $n$th term where $n = 107$. $l = 4(1.5)^{106}$.

With a scientific calculator:
$\log 4(1.5)^{106} = \log 4 + 106 \log 1.5$
$= 0.60206 + 106(0.1760913)$
$= 19.267733$
The antilog of 19.267733 (using the $10^x$ key) is $1.8524 \times 10^{19}$.

# *Lesson Eighty-nine*

## Sum of a Geometric Progression, Sigma Notation

Do the *Your Turn to Play*.

> *Life of Fred:*
> *Advanced Algebra*
> pp. 276–279

1. Find the sum of the first 20 terms of $1 + 0.9 + 0.9^2 + 0.9^3 + \ldots$.

2. Express in sigma notation: $6x^6 + 7x^7 + 8x^8 + \ldots + 40x^{40}$.

3. Express in sigma notation: $6x^7 + 7x^8 + 8x^9 + \ldots + 40x^{41}$.

4. Express in sigma notation: $9x^6 + 10x^7 + 11x^8 + \ldots + 43x^{40}$.

5. Fill in the question marks: $9x^6 + 10x^7 + 11x^8 + \ldots + 43x^{40} = \sum_{i=1}^{35} ?x^?$.

6. Fill in the question mark: $x^4 + x^5 + x^6 + \ldots + x^{124} = \sum_{i=4}^{?} x^i$.

7. How many terms are in the series $x^4 + x^5 + x^6 + \ldots + x^{124}$?

8. (You will need this for problem 3 in the city of Elizabeth which is coming up shortly.)  Suppose $-1 < r < 1$.

   Then $r$ might be any of these numbers:

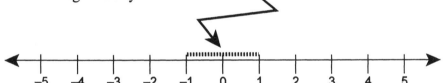

We want to look at what $r^n$ becomes as $n$ becomes very large.

Pick any value of $r$ so that $-1 < r < 1$.  You might choose, for example, $r = -0.3$ or $r = ¼$ or $r = 0.8$.

Use your calculator and begin computing the values of $r^2$, $r^3$, $r^4$, $\ldots$ and work until you can guess what $r^n$ will become when $n$ becomes very large.

---

### *Intermission*

In calculus, this computation will be called "finding the limit as n approaches infinity of $r^n$.
It is symbolized: $\lim_{n \to \infty} r^n$.

---

1. The first term, $a$, of $1 + 0.9 + 0.9^2 + 0.9^3 + \ldots$ is 1. The common ratio, $r$, is 0.9. We are looking for the sum of the first 20 terms so $n = 20$.

$$s = \frac{a(1 - r^n)}{1 - r} = \frac{1(1 - 0.9^{20})}{1 - 0.9} = \frac{1 - 0.9^{20}}{0.1}$$

If I wanted an approximate answer, I could use the $y^x$ key or logarithms. $0.9^{20} \approx 0.1215767$.

Then $s \approx \dfrac{1 - 0.1215767}{0.1} = 8.7842335$.

2. $6x^6 + 7x^7 + 8x^8 + \ldots + 40x^{40} = \displaystyle\sum_{i=6}^{40} ix^i$

3. $6x^7 + 7x^8 + 8x^9 + \ldots + 40x^{41} = \displaystyle\sum_{i=6}^{40} ix^{i+1}$

4. $9x^6 + 10x^7 + 11x^8 + \ldots + 43x^{40} = \displaystyle\sum_{i=6}^{40} (i+3)x^i$

5. $9x^6 + 10x^7 + 11x^8 + \ldots + 43x^{40} = \displaystyle\sum_{i=1}^{35} (i+8)x^{i+5}$.

Looking at the answers to questions 4 and 5, we note that there are several ways to express $9x^6 + 10x^7 + 11x^8 + \ldots + 43x^{40}$ in sigma notation.

6. $x^4 + x^5 + x^6 + \ldots + x^{124} = \displaystyle\sum_{i=4}^{124} x^i$.

7. At first blush* many people say that there are 120 terms. By that same reasoning, there would be 30 days in January.

8. I chose $r = 0.47$. Then $r^2 = 0.2209$, $r^3 = 0.103823$, $r^4 = 0.0487968$, $r^5 = 0.0229344$, $r^6 = 0.0107791$, and it looks like $r^n$ when $n$ gets very large is going to equal zero.

---

✱ *At first blush* is an idiom which entered our language about 1300. It means *upon first considering something.*

## Lesson Ninety

### End of the Chapter—Review & Testing
### Part One

---

Do all the problems in the first two cities.

*Life of Fred:*
*Advanced Algebra*

The Cities starting
on p. 280

Caldwell

Elizabeth

---
---

## Lesson Ninety-one

### End of the Chapter—Review & Testing
### Part Two

---

Do all the problems in
the second pair of cities.

*Life of Fred:*
*Advanced Algebra*

The Cities starting
on p. 281

Odd answers are in
the text, and even
answers are
given here.

 Unionville

 Valentine

---

### ✹E✹V✹E✹N✹ ✹A✹N✹S✹W✹E✹R✹S✹

Unionville

2. 860

4. 63

Valentine

2. 405

4. two rows, one column

Do all the problems in
the third pair of cities.

*Life of Fred:*
*Advanced Algebra*

The Cities starting
on p. 282

 York

Ideal

★A★N★S★W★E★R★S★

York

1. $-3(1 - 2^{30})$ or $3(2^{30} - 1) \approx 3.2212 \times 10^9 = 3,221,200,000$

2. 2565

3. 2482

4. $\dfrac{6(1 - (1/3)^6)}{1 - 1/3} \approx 8.988$

5. 9

6. F: harmonic;  G: geometric with $r = -1$;  H: geometric with $r = 0.1$;
I: arithmetic with $d = -4$;  J: none of the above

7. It is also a geometric progression (with the new common ratio equal to
the reciprocal of the original common ratio).

Ideal

1. 502,000

2. Matrix A must also have dimensions of 14 rows and 6 columns.

3. $(7 - y) + (7 - 2y) + (7 - 3y)$ or $21 - 6y$

4. 126 terms

5. Anything repeating such as $8 + 8 + 8 + 8 + 8 + \ldots$. This is an
arithmetic progression with $d = 0$ and a geometric progression with $r = 1$.
It's also an harmonic progression.

6. 1/2

7. The total cost of all the radio and TV ads placed in Cuba.

---

Do the *Your Turn to Play*.

```
Life of Fred:
Advanced Algebra
pp. 283–286
```

1. If Rita had her choice of 10 colors of jelly beans and 5 different outfits, how many different pictures could have been taken of her with every possible color of jelly bean and every possible outfit?

2. If you were writing a story and you could choose which one of the four bus tires would go flat, which one of the seven Dust Bunnies would be asked to go get a jelly bean, and which one of the remaining six Dust Bunnies would be asked to take the pictures, how many different stories could you have written?

3. The fundamental principle comes up in many different parts of mathematics. We used it when we asked the question (in Lesson Sixty-six) *If the domain has 35 elements and the codomain has 67 elements, how many different functions are possible?* Explain how the fundamental principle would lead us to the answer of $67^{35}$.

4. If the domain has 35 elements and the *range* has 67 elements, how many different functions are possible?

5. If the domain has 35 elements and the codomain has 67 elements, but the *range* has to have only 1 element, how many different functions are possible?

6. Suppose set A = {✳, Aunt Hilda, ❅}. It has three elements. Here are the **subsets**[*] of A: { }, {✳}, {Aunt Hilda}, {❅}, {✳, Aunt Hilda}, {✳, ❅}, {Aunt Hilda, ❅}, {✳, Aunt Hilda, ❅}.

Suppose set C = {$x_1$, $x_2$, $x_3$, . . . , $x_{35}$}. It has 35 elements. By the fundamental principle, determine how many possible subsets C has. This is not an especially easy problem. Expect to take 5–15 minutes to tease out the answer. *It is worth the effort.* Why? It's not really so important that you know how many subsets a 35-element set has. It is important that you strengthen your ability to reason analytically.

---

[*] The official definition: Set B is the **subset** of set A if every element in B is also in A.

# ANSWERS

1. By the fundamental principle, if there are 10 ways to do one thing (pick a color of jelly bean) and there are 5 ways to do another thing (pick an outfit), then there are 50 ways to do both.

2. 4 tires × 7 Dust Bunnies to get the jelly bean × 6 Dust Bunnies to take the pictures = 168 different stories.

3. Look at the first element in the domain. There are 67 possible images for it in the codomain. Then look at the second element in the domain. There are also 67 possible images for it. For each of the 35 elements in the domain, we have 67 possible choices. By the fundamental principle there are $35 \times 35 \times 35 \times 35 \times 35 \times \ldots \times 35 \times 35 \times 35 = 67^{35}$ possible ways to create a function which has 35 elements in its domain and 67 elements in its codomain.

4. The definition of the range of a function is the set of elements in the codomain that are images of at least one element of the domain. Since there are 35 elements in the domain and since each element of the domain has exactly one image, there can't be more than 35 images in the codomain. The range can't have more than 35 elements. There are no functions with a domain of 35 elements and a range of 67 elements.

5. Look at the first element of the domain. There are 67 possible images for it in the codomain. Then look at the second element in the domain. Since we are told that the range has only one element, the choice for the image of the second element of the domain is restricted to whatever element you chose for the image of the first element of the domain.

That seems like a lot of *words*. Let's see if it's any easier with symbols. Let $x_1$, $x_2$, $x_3$, . . . , $x_{35}$ be the elements of the domain. Let $y_1$, $y_2$, $y_3$, . . . , $y_{67}$ be the elements of the codomain.

Suppose you mapped $x_1$ to $y_7$. Then $x_2$ would also have to be mapped to $y_7$. And $x_3 \rightarrow y_7$ and . . . and $x_{35} \rightarrow y_7$.

By the fundamental principle, since there are 67 ways of mapping $x_1$ and only one way of mapping $x_2$, one way of mapping $x_3$, etc., there are $67 \times 1 \times 1 \times \ldots \times 1 = 67$ different possible functions.

6. If I want to specify a subset of set $C = \{x_1, x_2, x_3, \ldots, x_{35}\}$, I'm going to have to say whether $x_1$ is in that subset. I have two choices: Yes or No. After I've made that choice, I look at $x_2$ and ask the same question: will $x_2$ be in the subset. Again I have two possible alternatives: $x_2$ is in the subset or $x_2$ isn't in the subset. 35 decisions. Each with two alternatives. $2 \times 2 \times 2 \times 2 \times 2 \times \ldots \times 2 = 2^{35}$ possible subsets.

---

Do the *Your Turn to Play*. PROBLEM 5 SHOULD READ

P(12, 2) = ?

┌─────────────────────────┐
│ *Life of Fred:* │
│ *Advanced Algebra* │
│ pp. 287–289 │
└─────────────────────────┘

1. Fred had 18 different fountain pens. He wanted to pick one of them to write his trig lecture notes, a second one to write his calculus lecture notes, and a third one to write his topology lecture notes. Since each fountain pen was filled with a different color ink, he would then be able to keep his various notes separate. How many ways could he make his three choices?

2. Fred had received these 18 pens as a gift from the alumni association of KITTENS University in appreciation for the fine teaching that he had done in the first two years he had been there. After his third year of teaching they gave him a collection of 18 bottles of fountain pen ink, each a different color.

Then Fred decided to fill all 18 pens using a different color in each pen. How many ways could he do this?

3. Continuing the previous problem, suppose they had given him 40 different bottles of ink instead of just 18. How many ways could Fred have filled his 18 pens using a different color in each pen?

4. Continuing the previous problem, suppose that he had 40 different bottles of ink that he was going to use to fill his 18 pens, but he didn't care if two (or more) pens were filled with the same color. How many different ways could he fill his pens?

5. If the set of 18 pens were the domain and the 40 different inks were the codomain, how many possible functions could be created?

1. $P(18, 3) = \dfrac{18!}{15!} = \dfrac{18 \times 17 \times 16 \times 15 \times \ldots \times 3 \times 2 \times 1}{15 \times \ldots \times 3 \times 2 \times 1}$

$= 18 \times 17 \times 16 = 4896$ ways of choosing one pen out of the 18 for trig, a second pen for calculus and a third pen for topology.

2. He had 18 possibilities for the first pen. After that pen was filled, he had 17 alternatives to fill the second pen. All together he had 18! ways of choosing which color went into which pen.

3. There were 40 possibilities for the first pen, 39 for the second pen, 38 for the third pen. . . . All together he would have had $40 \times 39 \times 38 \times \ldots \times 25 \times 24 \times 23$ choices.

   A second way to approach this problem: Fred had to choose 18 bottles of ink out of the 40. $P(40, 18) = \dfrac{40!}{22!} = 40 \times 39 \times 38 \times \ldots \times 25 \times 24 \times 23$.

4. He had 40 choices for the first pen, 40 for the second, 40 for the third. By the fundamental principle, he had $40 \times 40 \times 40 \times \ldots \times 40 = 40^{18}$ possible ways of filling his pens.

5. This is exactly the same problem as the previous problem. For each pen (in the domain) he had to choose exactly one element in the codomain (exactly one color of ink). The number of functions from a domain with 18 elements to a codomain with 40 elements is $40^{18}$.

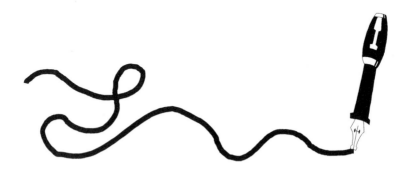

# *Lesson Ninety-five*

## Combination of n Things Taken r at a Time

---

1. $\dfrac{88!}{83!} = ?$

*Life of Fred:*
*Advanced Algebra*
pp. 290–292

2. $\dfrac{500!}{497!} = ?$

3. When the alumni association decided to give Fred 18 different pens, they went online and found that there were 2349 possible pens to choose among. The order in which the 18 pens were selected didn't matter. For the presentation ceremony the alumni association president handed Fred the 18 pens in a paper bag.

The formula for the number of ways of selecting r items out of a universe of n items where the order doesn't matter is

$$C(n, r) = \dfrac{n!}{(n - r)!\,r!}$$

How many ways could those 18 pens have been selected?

4. One member of the alumni association wanted to look at some of the 18 pens in the paper bag. He reached in and grabbed three of them at random and pulled them out to look at them. How many possibilities were there?

5. When the first member of the alumni association was done playing with the pens, he returned the 3 pens to the bag. A second member also wanted to look at some of the 18 pens in the paper bag. She reached in and grabbed 15 pens and pulled them out to look at them. How many possibilities were there?

6. Solve $\sqrt{x + 10} = x - 2$

Do the *Your Turn to Play*.

# ✸A✸N✸S✸W✸E✸R✸S✸

1. $\dfrac{88!}{83!} = \dfrac{88 \times 87 \times 86 \times 85 \times 84 \times 83 \times \ldots \times 2 \times 1}{83 \times \ldots \times 2 \times 1} =$

$= 88 \times 87 \times 86 \times 85 \times 84$

2. $\dfrac{500!}{497!} = 500 \times 499 \times 498$ (using the same write-it-out-and-cancel method used in the previous problem.)

3. $C(2349, 18) = \dfrac{2349!}{2331! \, 18!}$

If you wanted to work this out further, you could write

$\dfrac{2349!}{2331! \, 18!} = \dfrac{2349 \times 2348 \times 2347 \times \ldots \times 2332}{18 \times 17 \times 16 \times \ldots \times 2 \times 1}$

and then just multiply and divide on your calculator.

4. $C(18, 3) = \dfrac{18!}{15! \, 3!} = \dfrac{18 \times 17 \times 16}{3 \times 2 \times 1} = 816$

5. $C(18, 15) = \dfrac{18!}{3! \, 15!} = \dfrac{18 \times 17 \times 16}{3 \times 2 \times 1} = 816$

Why are the answers to these two problems the same?  Think of it this way: Every time she pulled 15 pens out of the bag, she was leaving 3 pens in the bag.  One way of identifying which 15 pens she pulled out of the bag would be by naming the 3 pens she left in the bag.  $C(18, 15) = C(18, 3)$.

6.             $\sqrt{x + 10} = x - 2$

Square both sides       $x + 10 = x^2 - 4x + 4$

Transpose             $0 = x^2 - 5x - 6$

Does it factor?  Yes.     $0 = (x - 6)(x + 1)$      (If it didn't factor, we'd have to use the quadratic formula.)

$$x - 6 = 0 \;\; OR \;\; x + 1 = 0$$
$$x = 6 \;\; OR \;\; x = -1$$

Ater you have squared both sides of an equation, it is mandatory to check each answer since you may have introduced extraneous roots.

Checking $x = 6 \ldots$             Checking $x = -1 \ldots$

$\sqrt{6 + 10} \overset{?}{=} 6 - 2$   Yes.        $\sqrt{-1 + 10} \overset{?}{=} -3$   No.

# *Lesson Ninety-six*

## The Binomial Formula

---

Do the *Your Turn to Play.*

*Life of Fred:*
*Advanced Algebra*
pp. 293–298

Here is the pattern for writing out the first several terms of $(x + y)^{26}$ :

$$(x + y)^{26} = x^{26} + \frac{26}{1!} x^{25}y + \frac{26 \cdot 25}{2!} x^{24}y^2 + \frac{26 \cdot 25 \cdot 24}{3!} x^{23}y^3 + \ldots$$

1. Write the first three terms of the binomial expansion of $(x + y)^{55}$.

2. Give the first four terms of $(w + 4)^{21}$.

3. The first three terms of $(x^3 + 2y)^{800}$.

4. $w^{45} + 45w^{44}z + \dfrac{45 \cdot 44}{2!} w^{43}z^2$ are the first three terms of the expansion of what binomial?

5. What is the twenty-second term of the expansion of $(x + 5y^8)^{4000}$ ?

6. This is the seventh term in a binomial expansion. Fill in the value of the question mark: $\dfrac{18 \cdot 17 \cdot 16 \cdot 15 \cdot 14 \cdot 13}{6!} x^? y^6$

7. How many terms are in the expansion of $(3x + 44y)^{300}$?

8. What is the equation of the line that passes through (3, 4) and (–2, 7)?
   (Hint: All the formulas for lines are found on p. 130.)

1. $(x + y)^{55} = x^{55} + 55x^{54}y + \dfrac{55 \cdot 54}{2!} x^{53}y^2 + \ldots$

2. $(w + 4)^{21} = w^{21} + 21w^{20}(4) + \dfrac{21 \cdot 20}{2!} w^{19}(4)^2 + \dfrac{21 \cdot 20 \cdot 19}{3!} w^{18}(4)^3 + \ldots$

   $= w^{21} + 84w^{20} + 3360w^{19} + 85120w^{18} + \ldots$

3. $(x^3 + 2y)^{800} = (x^3)^{800} + 800\,(x^3)^{799}(2y) + \dfrac{800 \cdot 799}{2!} (x^3)^{798}(2y)^2 + \ldots$

   $= x^{2400} + 1600x^{2397}y + 1278400x^{2394}y^2 + \ldots$

4. $(w + z)^{45}$

5. The $22^{nd}$ term of the expansion of $(x + 5y^8)^{4000}$ is

$$\dfrac{4000 \cdot 3999 \cdot \ldots \cdot 3980}{21!} x^{3979}(5y^8)^{21}$$

6. $\dfrac{18 \cdot 17 \cdot 16 \cdot 15 \cdot 14 \cdot 13}{6!} x^{12}y^6$

7. We first note that $(x + y)^0 = 1$ has one term.

   $(x + y)^1 = x + y$ has two terms.

   $(x + y)^2 = x^2 + 2xy + y^2$ has three terms.

   $(x + y)^3 = x^3 + 3x^2y + 3xy^2 + y^3$ has four terms.

$(3x + 44y)^{300}$ has 301 terms.

8. Using the two-point form of the line $\dfrac{y - y_1}{x - x_1} = \dfrac{y_2 - y_1}{x_2 - x_1}$

the equation through (3, 4) and (−2, 7) is $\dfrac{y - 4}{x - 3} = \dfrac{7 - 4}{-2 - 3}$

$$\dfrac{y - 4}{x - 3} = \dfrac{3}{-5}$$

## *Lesson Ninety-seven*
### Pascal's Triangle

Just read and enjoy the end of the story.

*Life of Fred:*
*Advanced Algebra*
pp. 299–302

## *Lesson Ninety-eight*
### End of the Chapter—Review & Testing
### Part One

Do all the problems in the first two cities.

 Calhoun

 Park City

*Life of Fred:*
*Advanced Algebra*
The Cities starting
on p. 303

## *Lesson Ninety-nine*
### End of the Chapter—Review & Testing
### Part Two

Do all the problems in
the second pair of cities.

Odd answers are in
the text, and even
answers are
given here.

 Walcott

 Zap

*Life of Fred:*
*Advanced Algebra*
The Cities starting
on p. 304

### ✸E✸V✸E✸N✸ ✸A✸N✸S✸W✸E✸R✸S✸

Walcott

2. Three letters offer $26^3 = 17,576$ possibilities. Four digits offer $10^4 = 10,000$ possibilities.

4. First, five of the ten people are selected to receive oranges, C(10, 5). Then two of the remaining five are selected to receive bananas, C(5, 2). The rest automatically get mangos, C(3, 3) = 1. By the fundamental principle, the handing out of the oranges followed by the handing out of the bananas, followed by the handing out of the mangos =

C(10, 5) × C(5, 2) × C(3, 3) = $\frac{10!}{5!5!}$ × $\frac{5!}{3!2!}$ × $\frac{3!}{0!3!}$ = 2520

Zap

2. 1  11  55  165  330  462  462  330  165  55  11  1

4. Fredrika had four choices. After she made her choice, then Meddie had three possible choices and then Rita had two choices. By the fundamental principle, the total possible outcomes was 4 × 3 × 2 = 24.

## *Lesson One hundred*

### End of the Chapter—Review & Testing
### Part Three

Do all the problems in
the third pair of cities.

*Life of Fred:*
*Advanced Algebra*

The Cities starting
on p. 305

 Dallas

 Independence

## ✸A✸N✸S✸W✸E✸R✸S✸

Dallas

1. 6!  or  720

2. P(12, 3)  or  12·11·10  or  1320

3. $x^{44} + 44x^{43}y + \frac{44 \cdot 43}{2!} x^{42}y^2 + \frac{44 \cdot 43 \cdot 42}{3!} x^{41}y^3$

4. C(6, 2) × P(4, 4)  or  $\frac{6!}{4!2!}$ × 4! = 360

5. There were two choices on the first line and two on the second and two on the third, etc. By the fundamental principle, there were $2 \times 2 \times 2 \times 2 \times 2 \times 2 = 2^6 = 64$ possible way to fill out the form.

Independence

1. $C(50, 4) \times C(60, 4) \times C(90, 4)$ or $\dfrac{50!}{46!4!} \times \dfrac{60!}{56!4!} \times \dfrac{90!}{86!4!}$

2. $14^{100}$ This is the fundamental principle. There are 14 possible images for the first element of the domain. There are 14 possible images for the second element of the domain. Etc.

3. Using the 6th row of Pascal's Triangle: 1 6 15 20 etc., we have 6 out of a total of $2^6$ possibilities $= \dfrac{6}{64}$ or $\dfrac{3}{32}$

4. 31! which is approximately $8.2228 \times 10^{33}$

5. $C(18, 5) = \dfrac{18!}{13!5!} = 8568$

$\mathscr{Lesson\ One\ hundred\ and\ one}$

The Hardest Problem in Advanced Algebra

*Life of Fred:*
*Advanced Algebra*
pp. 306

Some problems are *long*. Some problems are *hard*.

If I'm being paid by the hour, I love *long* problems. Pay me a decent wage and I'll be glad to trim your lawn blade-by-blade using a pair of tweezers. I'll bake 400 dozen cupcakes.

Some math books seems to specialize in long problem sets. They might tell you that the area of a circle is $A = \pi r^2$ and then give you forty problems like: 1. $r = 5$, $A = ?$ 2. $r = 6$, $A = ?$ 3. $r = 2$, $A = ?$ 4. $r = 9$, $A = ?$ 5. $r = 7$, $A = ?$ 6. $r = 1$, $A = ?$ 7. $r = 10$, $A = ?$, etc. You put your mind on GRIND mode and think about what kind of pizza you are going to have tonight as you plow through the forty problems.

But it's the *hard* problems that offer the fun. They ask for insight rather than for repetitive motion.

So you are given a dozen Waddle's Doughnuts. They come in six flavors: $A_{pple}$, $B_{lueberry}$, $C_{hocolate}$, $D_{usted}$, $E_{ggplant}$, and $F_{rosted}$. They're tossed into a big white bag and handed to you. You are asked how many different possibilities are there.

If you look at it right, there is very little computation that you have to do. Out will pop the answer of 6188 possibilities. (That is the answer.)

But this problem has lots of blind alleys. Lots of approaches that don't work. That is, by the way, the definition of a *hard* problem.

When I first tried to work the problem, I started by wondering how many Apple doughnuts would go into the white bag. There were 13 possible cases. I could have had zero Apple doughnuts or I could have had 12. Then for each of those 13 possible cases, I thought about how many possible Blueberry doughnuts. For example, in the case I had 4 As, then I could have had anywhere from zero to 8 Bs. That's nine subcases. Too much.

In another approach I imagined that there were 12 slots in a row in which to stick the doughnuts: _ _ _ _ _ _ _ _ _ _ _ _
There were 6 possible choices for the first slot, 6 for the second, etc. and so by the fundamental principle there would be $6 \times 6 \times 6 \times 6 \times 6 \times 6 \times 6 \times 6 \times 6 \times 6 \times 6 \times 6 = 6^{12}$ possible ways to fill those slots. But in the problem the doughnuts are all tossed into a white bag. The order doesn't matter. Among the $6^{12}$ arrangements we would have all of the following:

        AAABBBCCCCCC
        AABABBCCCCCC
        AABBBCCCCCCA
        ABBBAACCCCCC    but these are all the same Waddle's Special.
I was stuck.

On the next page is the solution I found.

Imagine when you get home with your big white Waddle's Special bag, that you open it up and line up the doughnuts to see how many of each kind you've selected.

First line up the Apple doughnuts (like Lifesavers) on edge. There might be any number of As from zero to twelve. Then after the As put a black rubber doughnut.

Then put the Bs in the line and then put another black rubber doughnut. Then the Cs and another black rubber doughnut and then the Ds and another black rubber doughnut and the Es and another black rubber doughnut and then, finally, the Fs.

I've put five black rubber doughnuts--sort of like bookmarks--between the various kinds of doughnuts.

If I had 2As, 5Bs, 0Cs, 1D, 1E and 3Fs, it would look like: AA**R**BBBBB**RR**D**R**E**R**FFF where **R** stands for the black **R**ubber doughnuts.

If I had 12Es, it would look like **RRRR**EEEEEEEEEEEE**R**.

In any case, we're looking at 17 slots where 5 of them are filled with **R**s and the rest with the six kinds of doughnuts.

Now suppose we carefully remove the edible doughnuts and just leave the spaces they occupied and the **R**s. From my first example, AA**R**BBBBB**RR**D**R**E**R**FFF would become

$$\_\ \_\ \textbf{R}\ \_\ \_\ \_\ \_\ \_\ \textbf{R}\textbf{R}\ \_\ \textbf{R}\ \_\ \textbf{R}\ \_\ \_\ \_\ .$$

From my second example, **RRRR**EEEEEEEEEEEE**R** would become

$$\textbf{RRRR}\_\ \_\ \_\ \_\ \_\ \_\ \_\ \_\ \_\ \_\ \_\ \_\ \textbf{R}$$

Now the thing to notice is that if you give me any 17 slots with five of them filled in with **R**s, you have uniquely specified a Waddle's Special. And, likewise, if you had a Waddle's Special it will uniquely specify 17 slots with five slots filled in with **R**s.

(In the words of the chapter on functions, there is a 1-1 correspondence between the set of all Waddle's Specials and the set of all 17 slots with five of them filled in with **R**s. And we know that if two sets can be placed into 1-1 correspondence with each other, they have the same number of elements.)

The rest is easy. How many ways can you pick five of the 17 slots (to put the **R**s in)? That C(17, 5). We're done.

$$C(17, 5) = \frac{17!}{12! \, 5!} = 6188$$

# Index

# Polka Dot Publishing

## Life of Fred: Beginning Algebra

Numbers, Integers, Equations, Motion & Mixture, Two Unknowns, Exponents, Factoring, Fractions, Square Roots, Quadratic Equations, Functions & Slope, Inequalities & Absolute Value.
ISBN: 0-9709995-1-8, 320 pages. $29

## Fred's Home Companion: Beginning Algebra

*Life of Fred: Beginning Algebra* divided into daily lessons.
*Life of Fred: Beginning Algebra* answer key.
Many more beginning algebra problems.
ISBN: 0-9709995-6-9, 128 pages. $14

## Life of Fred: Advanced Algebra

Ratio, Proportion & Variation, Radicals, Logarithms, Graphing, Systems of Equations, Conics, Functions, Linear Programming, Partial Fractions, Math Induction, Sequences, Series, Matrices, Permutations & Combinations.
ISBN: 0-9709995-2-6, 320 pages. $29

## Fred's Home Companion: Advanced Algebra

*Life of Fred: Advanced Algebra* divided into daily lessons.
*Life of Fred: Advanced Algebra* answer key.
Many more advanced algebra problems.
ISBN: 0-9709995-7-7, 128 pages. $14

## Life of Fred: Geometry

Points and Lines, Angles, Triangles, Parallel Lines, Perpendicular Lines, Quadrilaterals, Area, Similar Triangles, Symbolic Logic, Right Triangles, Circles, Constructions, Non-Euclidean Geometry, Solid Geometry, Geometry in Four Dimensions, Coordinate Geometry, Flawless (Modern) Geometry.
ISBN: 0-9709995-4-2, 544 pages. $39

## Life of Fred: Trigonometry

Sines, Cosines and Tangents, Graphing, Significant Digits, Trig Functions of Any Angle, Trig Identities, Graphing $a \sin (bx + c)$, Radian Measurement, Conditional Trig Equations, Functions of Two Angles, Oblique Triangles, Inverse Trig Functions, Polar Coordinates, Polar Form of Complex Numbers, Preview of all of Calculus.
ISBN: 0-9709995-3-4, 320 pages. $29

## Fred's Home Companion: Trigonometry

*Life of Fred: Trigonometry* divided into daily lessons.
*Life of Fred: Trigonometry* answer key.
Many more trigonometry problems.
ISBN: 0-9709995-8-5, 128 pages. $14

## Life of Fred: Calculus

Functions, Limits, Speed, Slope, Derivatives, Concavity, Trig, Related Rates, Curvature, Integrals, Area, Work, Centroids, Logs, Conics, Infinite Series, Solids of Revolution, Polar Coordinates, Hyperbolic Trig, Vectors, Partial Derivatives, Double Integrals, Vector Calculus, Differential Equations.
ISBN: 0-9709995-0-X, 544 pages. $39

## Life of Fred: Statistics

Descriptive Statistics (averages, measures of dispersion, types of distributions), Probability, Bayes' Theorem, From a Given a Population Determine What Samples Will Look Like (7 tests), Techniques of Sampling, From a Given Sample Determine What the Population Was (14 tests), Determine Whether Two Given Samples Came From the Same Population (15 tests), Working With Three or More Samples (10 tests), Emergency Statistics Guide, Regression Equations, Field Guide, 16 Tables.
ISBN: 0-9709995-5-0, 544 pages. $39

## Order Form

_____ copies of: *Life of Fred: Beginning Algebra*                    $29  $_____

_____ copies of: *Fred's Home Companion:*
                            *Beginning Algebra*                    $14  $_____

_____ copies of: *Life of Fred: Advanced Algebra*                    $29  $_____

_____ copies of: *Fred's Home Companion:*
                            *Advanced Algebra*                    $14  $_____

_____ copies of: *Life of Fred: Geometry*                    $39  $_____

_____ copies of: *Life of Fred: Trigonometry*                    $29  $_____

_____ copies of: *Fred's Home Companion:*
                            *Trigonometry*                    $14  $_____

_____ copies of: *Life of Fred: Calculus*                    $39  $_____

_____ copies of: *Life of Fred: Statistics*                    $39  $_____

Shipping & Handling        $___free___

Total enclosed        $_____

*Send the books to:*

YOUR NAME_____

ADDRESS_____

CITY STATE ZIP_____

Mail this order with your check or money order to:
Polka Dot Publishing
P. O. Box 8458
Reno NV 89507–8458

If you would like to visit Fred at his official
Web site and see a complete list of the books that have been
written about him, go to    FredGauss.com


At the Web site you can also order by credit card.